The Software Project Rosetta Stone
Use Case Analysis

The Software Project Rosetta Stone

Use Case Analysis

David A. Bly

CreateSpace
Seattle, WA

CreateSpace Independent Publishing Platform

Copyright © 2014 by David A. Bly

All rights reserved. No part of this book may be reproduced or transmitted in any form or by any means, electronic or mechanical, including photocopying, recording, or by any information storage and retrieval system, without written permission from the author, except for the inclusion of brief, attributed quotations in a review.

Cover image: "Rosetta Stone" by Hans Hillewaert — Own work. Licensed under Creative Commons Attribution — Share Alike 3.0 via Wikimedia Commons. http://commons.wikimedia.org/wiki/File:Rosetta_Stone.JPG#mediaviewer/File:Rosetta_Stone.JPG

International Standard Book Numbers

ISBN-13: 978-1503197589
ISBN-10: 1503197581

Printed in the United States of America

Dedication

To Annie, my far better half, who has provided me with constant and consistent support.

Preface

Clear definitions of project scope and functional requirements are two of the most fundamental contributors to project success. The opposite is also true. Unclear definitions of project scope and functional requirements are two of the most common reasons for project failure. I am surprised to see how often these problems occur, especially since it does not have to be that way.

I have found that use cases provide an excellent mechanism to achieve clarity on both counts. They provide an unequivocal definition of functional scope and they define functional requirements in ways that add value to the work of all your team members. Further, they help you track the necessary capabilities from initial requirements through all subsequent project work: design, development, system test, user acceptance, training, and support. Finally, they serve as a Rosetta Stone for your entire project team by providing a unified understanding of the system for the various perspectives that contribute to the project's success.

This little book presents a brief introduction to use cases and shows you how to employ them to develop interactive software. Widely acknowledged in the industry, they can take many different forms and formats. This book presents one view. In practice, you and your project team will want to tailor use cases to suit your needs and collective preference.

I would like to acknowledge several individuals who have contributed to this endeavor. First, I would like to thank those industry sages who invented use cases and showed us how they can help. Ivar Jacobson is credited with first popularizing the concepts of use case analysis. He describes them in his book, "Object-Oriented Software Engineering: A Use Case Driven Approach" (Jacobson, et al. 1992). Several others helped to drive out use case conventions as well, such as Gunnar Övergaard. More recently, Alistair Cockburn has written an excellent book entitled "Writing Effective Use Cases" (Cockburn 2001), which has guided many of us to new heights in our practice of use case analysis.

I would also like to thank several people who have taught me to manage projects better and helped me in my career over the course of thirty years. Rick Roller, the best manager I ever had, helped me shape my professional ethic. Bill Browne, my first boss in the consulting industry, trusted me to serve my clients the way I wanted to but was always there as my safety net if I needed it. Tish Hill first asked me to mentor project managers and to develop my first training course in software project management. Ross Porter first educated me on use cases. Ken Schwaber taught my Certified Scrum Master training course and raised my level of thought about software development. Tim Henebry trusted me with a monster project and showed me what effective project sponsorship looks like.

I would also like to thank my friends who encouraged and assisted me in the creation of this book. Longtime friends Eli Embley, Steven Salta, and Van Chesnutt spent hours reviewing this text and offered many helpful suggestions. Matt Simon, an excellent writer and friend, labored over my drafts to help rid me of my writing demons. And, Dale Carlson, another long-time, esteemed colleague, designed the covers for this book. I am extremely grateful for all of their assistance.

TABLE OF CONTENTS

Introduction 1
   About This Book 3
   Benefits of Reading this Book 5
   Audience 6
   My Goal 7

What are Use Cases? 9
   What Do Use Cases Do? 9
   Key Terms 11
   Why Use Cases? 12
   Forms of Use Cases 19
   Use Cases Compared to Business Process Modeling 26
   Use Cases Compared to User Stories 29
   When to Employ Use Cases 34
   Other Considerations 39

Use Case Roles and Responsibilities 41
   What is a Role? 41
   The Roles in Use Case Analysis 42
   Responsibility Matrix 45

Developing Use Cases 49
   Precision Levels 52
   Use Case Discovery 55
   Initial Use Case Definition 65
   Use Case Elaboration 85
   Pitfalls to Avoid 91

Planning for Use Case Analysis **97**
   Use Cases in Project Commissioning 98
   Use Cases in Waterfall Project Delivery 101
   Use Cases in Agile Project Delivery 104
   Project Planning 110
   Prioritizing Use Cases and Stories 114
   Use Cases in Project Closure 117
   Use Cases in Application Support 118

In Summary **119**

References **121**

Appendix A – Use Case Template **123**

Appendix B – Glossary **127**

Index **137**

Table of Figures

Figure 1: Describing Use Cases 10
Figure 2: Why Use Cases? 15
Figure 3: Basic Artifact Map 17
Figure 4: Use Case Diagrams 21
Figure 5: Fully Dressed Use Case Example 23
Figure 6: UML Sequence Diagram 25
Figure 7: Flow Chart 26
Figure 8: Workflow Model Example 27
Figure 9: Vertical Slices 30
Figure 10: Use Cases into Stories 31
Figure 11: Use Case Applicability 36
Figure 12: Role Matrix 42
Figure 13: Responsibility Matrix 46
Figure 14: The Cone of Uncertainty 50
Figure 15: Precision Levels 53
Figure 16: A Fully Dressed Use Case 54
Figure 17: Use Case Register 63
Figure 18: Sample Use Case – Precision Level 0 – Discovery 69
Figure 19: Sample Use Case – Precision Level 1 – Scope 69
Figure 20: Sample Use Case – Precision Level 2 – Main Success Scenario 73
Figure 21: Page-Flow Example 77
Figure 22: Page Mockup Example 78
Figure 23: Requirements Register 79
Figure 24: Business Rule Register 82
Figure 25: Project Issue Log 84
Figure 26: Fully Dressed Tool Room Use Case 88

Figure 27: Tool Room User Stories 90
Figure 28: A Generic Project Life Cycle Model 97
Figure 29: High-Level Estimating with Use Cases 99
Figure 30: High-Level Estimating with Use Case Complexity 100
Figure 31: Waterfall Life Cycle 102
Figure 32: Waterfall Artifact Map 103
Figure 33: Pipelining Use Case Development and Testing 105
Figure 34: Agile Artifact Workflow 108
Figure 35: The "Done List" 109
Figure 36: Gantt Chart 110
Figure 37: Value-Driven Prioritization 116

INTRODUCTION

Coaching project managers on broken projects starts with troubleshooting the situation for what went wrong. Having engaged in many such sessions over the years, I am surprised how often the root cause of distress points back to scope and requirements definition.

When I ask to see the scope definition of the project, I often get back a short list of the goals, objectives, and required results of the project. There is not much there that would help me estimate the cost of the project or define the solution strategy to achieve the results. As a result, the statement of scope can be interpreted multiple ways. This ambiguity can lead to different expectations by both the customer and the producer. These varied expectations can lead to significant conflict if the project goes awry.

Scope must be defined in clear, unequivocal language that minimizes the possibility of different interpretations. That requires more discovery effort up front. The good news is that a little effort to define the use cases goes a long way. I like to call it Use Case Discovery. A brief workshop with the customer in the discovery phase to define the actors and their goals provides a definitive list of the capabilities to be built in the project and provides a reasonable basis for estimating the costs and schedules for the project.

> **Scope must be defined in clear, unequivocal language that minimizes the possibility of different interpretations.**

When troubleshooting a distressed project, I also ask to see the requirements that were defined for the project. In these agile days, a question like this can lead to debates about methodology (more about that later), but generally, if requirements have been defined at all, I get a list of statements about what the system must be able to do or how fast it should run.

It's good to list the things a system must be able to do, but I think it falls short. A mere list does not provide a bridge to the downstream steps in the Software Development Life Cycle (SDLC). The requirements must be written in such a way that they can be used to design an appropriate solution to the customers' needs. They must provide useful input to developers, testers, and documenters. I find that there is often a missing "miracle happens here" step in the project plan that leaps from a list of requirements to system design. This missing link results in poor system design, poor fit for the stated purpose, expensive rework, project overruns, and poor system acceptance.

> *... there is often a missing "miracle happens here" step in the project plan that leaps from a list of requirements to system design.*

Once again, I have found that use case analysis provides a ready solution to these problems. Use cases composed in cross-functional workshops document the functional requirements of the system in useful ways. User interface designers draw upon them to develop user interface mock-ups and wireframes. User education specialists base their documentation and training materials on them. Testers write their test cases for the scenarios contained in them. Developers infer application logic, class models and interfaces from them. Customers find assurance in them that the system will meet their needs. They serve as a Rosetta Stone for the many skill sets involved in today's multi-faceted system development.

In contrast to the documentation-heavy contracts disparaged in the *Agile Manifesto* (Agile Manifesto 2001), use cases also support test-driven development at the functional test level. A fully dressed use case very much embodies several functional test cases, one for each of the usage scenarios it contains. Furthermore, they can be developed incrementally in a project life cycle that balances waterfall and agile approaches. I like to think of them as a "Little Design Up Front."

Conveniently, the requirements statements about the system capabilities are a handy by-product of the use case analysis process. They arise in the course of use case discussions and can be captured in the moment in a separate list that can be refined later in the context of the use case scenarios. I will discuss this more in the remainder of this book.

About This Book

This book is intentionally small. It has grown out of a white paper I wrote several years ago. In past projects, I gave that paper to customers before we started a project together. It discussed use case analysis and how we would employ that on our project. It served its purpose well with an overview of use cases, an explanation of their benefits, and an illustration of their fit into the overall project.

I would like to think that this little book can serve a similar purpose for you. Hopefully, it will be useful for you and your team members as you embark on a project that will employ use cases or in situations where you are considering use cases. I have sought to define the basics and to identify how they fit into the larger whole of a software project without an exhaustive treatment of the subject.

There are many fine books available that will provide you with broad and deep coverage of use cases. Check the References section at the back end of this book for a list. These books provide excellent guidance if you need an in-depth education. However, they can be overkill for you and your project team members who will participate in but not lead the analysis.

> *I like to think of this book as the glass of water without the hose.*

That is where this little book applies. You might have heard the proverbial phrase, "It's like trying to get a glass of water out of a fire hose." I like to think of this book as the glass of water without the hose.

Samples Used

Several examples of use cases in their various flavors are provided in this book. I have intentionally varied the applications they represent so you can see how they fit multiple situations.

Contents

This book is divided into four main sections, each of which builds on the explanations provided in the section before it.

What are Use Cases?

This section provides you with a basic introduction to use cases, key terminology, and samples of the many forms use cases can take. You will learn about the relationships between use cases, business process models, and user stories. Finally, you will read about the appropriateness of use cases for different kinds of software projects.

Use Case Roles and Responsibilities
Several disciplines are involved in the definition and usage of the information developed in use case analysis. In this section, you will read about some typical roles and see how they are involved.

Developing Use Cases
Use case analysis begins in the discovery phase before the project is even launched. From there, you will learn how a progressive elaboration takes place throughout the project that includes user analysis, events, requirements, business rules, and issues.

Planning for Use Cases
This section offers some insights into the ways you can fit use case analysis into both waterfall and agile project plans. You will review some of the artifacts produced in use case analysis. Finally, you will read about ways you can prioritize use cases and project work based on business value.

Benefits of Reading this Book

After you read this book, I hope you will begin to enjoy the following benefits as you put to work the principles and practices you will learn here.

Improved Use Case Analysis
After reading this book, you will see ways to improve your implementation of use case analysis. You will make better decisions with your teams about the different forms and formats of use cases and which to employ. You will also be more aware of the project

types that are best suited for use case analysis and which project types do not benefit as much.

Increased project success

This book will provide you with tools and techniques to better set up your projects for success, to more accurately estimate costs, to better price your projects, and to deliver the full desired scope within the original cost and schedule estimates you produce.

Increased customer satisfaction

The guidelines provided in this book will enable you to produce computer systems that better fit your customers' desires and better serve their needs. Their satisfaction will increase as a result.

Increased team member satisfaction

When you apply what you learn in this book, cross-functional team members will feel more confident that their voices have been heard. They will understand the system better and will be better able to develop a system that meets all of the needs of the users.

Reduced project risk

Projects are risky business. Many things can go wrong. You will see that use case analysis can reduce risk through improved communications and handoffs among the team members and through better functional fit of the system for the customer.

Audience

You are my primary audience if you lead software projects, represent your business owners as a product owner, or otherwise directly contribute to the development of interactive software systems.

- As a project manager, you will learn to better orchestrate project activities to achieve project success.
- As a product manager, you will learn how to better define functional needs through use case analysis.
- As a business analyst, you will learn better tools to express processes and requirements.
- As a developer, you will better comprehend the analysis that defines the systems you will build.
- As a usability analyst, you will gain a better context to produce user interface designs and specifications.
- As a quality assurance analyst, you will be able to better participate in the early solution definition work to build quality in from the start and then test the results as they are produced.

However, other roles and functions will also benefit from reading this book. You should consider reading this book if you are a stakeholder on projects as a functional manager, a project office manager, or a project sponsor. It will give you a good overview of how software project teams figure out what to build.

My Goal

In my own work, I live by the very simple motto, "Add value." I succeed by helping my customers succeed. I strive to add value to my customers' agendas, not fit them into mine. I often find that the value I can add for my customers may not be recognized by either of us in my initial assignment, but as we get to know each other, we both see additional ways I can help.

I have applied that same philosophy to this book. I hope this book adds value for you. I want to help you be a better project manager, product owner, business analyst, architect, or other team member. If reading this book helps you to do that and to make your customers happy, I will have achieved my goal.

What are Use Cases?

Use Cases provide a way to capture the users' functional requirements for a computer system. In many circles, they are considered a best practice in software engineering. Use cases are recommended whenever there are significant conversations between the user and the system. They help design online web applications nicely because they keep the work centered on the user's needs. You begin defining them very early in the project and add more detail to them as the project progresses.

What Do Use Cases Do?

A single use case documents a conversation between an actor and a system to achieve a meaningful goal. An actor can be a person, a role played by one or more persons, an organization, or a computer system. These conversations have a give-and-take flow to them of request and response. It is this interaction that the use case captures as shown in Figure 1.

> *A use case documents a conversation between an actor and a system to achieve a meaningful goal.*

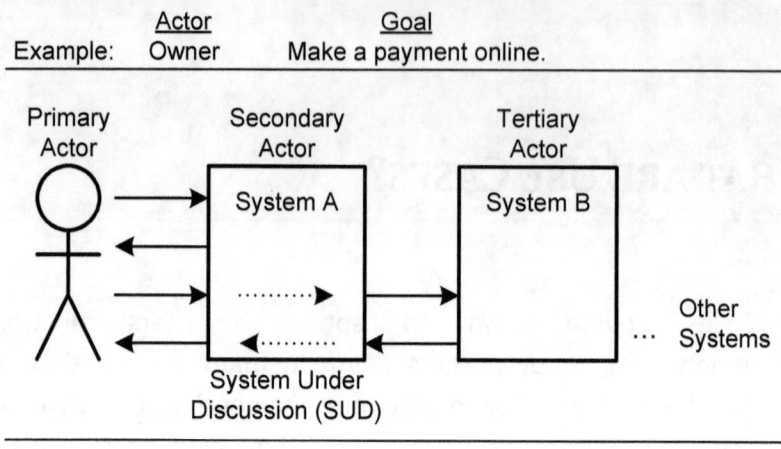

Figure 1: Describing Use Cases

Each use case documents the conversation necessary for one actor to achieve one meaningful goal. The goal must have some importance to the actor. When completed, the use case also documents the several different paths through which the conversation might flow, anticipates problems that might occur, and defines responses to those problems.

Obviously, an actor can have many goals for interacting with a computer system. Each goal is documented as a separate use case. Also, a computer system often has more than one actor or type of user. Each actor has a combination of unique goals and goals that are shared with other actors. Therefore, it is common to build a library of use cases to define the functional requirements for a system. It is considered best practice to maintain this library of use cases throughout the useful life of the system. They can be modified as needed to reflect changes in the needs of the user community over time. They can be supplemented as the

> *It is considered best practice to maintain the library of use cases throughout the useful life of the system.*

system capabilities grow, deleted when they are no longer needed, or merged with others.

Collectively, this library of use cases, referred to as a use case model, gathers together into one package all of the ways in which the system is used and the capabilities it provides to its using community (Övergaard and Palmkvist 2005).

Key Terms

Here are some key terms that you will encounter as you read through the remainder of this book.

- An *actor* initiates the use case to achieve a *goal*. The *primary actor* initiates the use case. *Secondary actors* respond to the primary actor or carry the conversation forward to other actors. The use case is complete when the conversation is concluded, even if the goal has not been achieved.
- The *System Under Discussion* (SUD), often a secondary actor, responds to the primary actor's actions and initiates conversations with other systems as needed to achieve the primary actor's goal (Cockburn 2001).
- The *Main Success Scenario (MSS),* sometimes referred to as the *Sunny Day Scenario, Normal Course, Basic Course, Happy Path, or Flow of Events,* documents the most straightforward, successful path of the conversation in a step-by-step fashion (Cockburn 2001).
- *Alternate Paths* document variations or forks off of this main path to address different scenarios in decision logic.
- *Exceptions* document anticipated error conditions and the responses to those errors.

- *Scenarios* represent the unique paths that can be followed through a use case.

I will discuss each of these terms further in this book.

Why Use Cases?

Use cases increase the development team's collective comprehension of the desired capabilities of the software they are about to build. You can employ them to reduce or eliminate several common problems associated with projects that develop or modify software for people.

The Voice of the User

Because they are user-centric, use cases keep your focus and the team's focus on the needs of the user and the user's experience.

Too often, software teams express requirements in the voice of the system. For example, "The xyz system must be able to apply discounts to the purchase price." It's almost as if the system is the center around which secondary users revolve. This is an antiquated manner of expression that harkens back to the days of master control programs and second-generation languages. It leads you to inconsiderate system designs where the needs of the system outweigh the needs of your users.

> *... use cases keep your focus and the team's focus on the needs of the user and the user's experience.*

Instead, consider that a system exists to assist its users. In this light, the voice of the user reigns. "What do you want to do, Mr. User? How can I help you?" This perspective leads you to design a system that people will want to adopt because it helps them solve their problems or makes their jobs easier.

Furthermore, use case analysis puts users' requirements within the context of the larger conversation (Bittner and Spence 2003). Requirements will make much more sense to you when they are understood as part of the accomplishment of meaningful results.

Translating Needs into Solutions

Ask the prospective users of a computer system what they need and they will give you a list of bullets that enumerate things the system must be able to do. This is a good start but a list of users' needs is not enough to design a computer system. It doesn't tell you how the system is expected to behave or how it should get from the start to the finish of a task or a series of tasks.

Without this extra information, developers are inclined to push forward to system solutions without a full comprehension of the problems (Leffingwell and Widrig 2003). Furthermore, they develop the system based on their own perspective of the users' needs, which may be very different from what a user or designer may visualize. As a result, user satisfaction and system adoption often suffer.

Use cases help you bridge the gap between users' needs and expected system behavior. The collaboration between users, designers, and developers to create the use cases provides better information to design and develop the system. They reinforce the need to define systems from the users' perspectives. This leads you to design a system that better fits the users' expectations.

Cross-Functional Understanding

Use case analysis is inherently cross-functional. When you define use cases you should include representatives from the user community, business owners, user-interface designers, business analysts, developers, and quality assurance analysts. Documenters and support analysts can also lend a valuable perspective.

The different perspectives of this cross-functional team emerge when you get them together in a room to discuss use cases. These differences are immediately made known to the other functions represented in the group. Decisions must make sense to everyone before being incorporated into the use case, so many differences are resolved through discussion as the use case is created. This integrated approach results in better overall designs. You have no "throw it over the wall" phenomenon associated with traditional requirements documents. Differences, points of confusion, and issues are surfaced and resolved immediately.

The famous Rosetta Stone, carved in 196 BCE, expressed the same message in three different languages and thereby provided crucial clues to deciphering ancient Egyptian hieroglyphs. The same benefits accrue to teams that employ use cases. The collection of use cases serves as a Rosetta Stone for the team. That is, it serves as the point of common understanding for all of the major perspectives in the development of software as illustrated in Figure 2.

> **The collection of use cases serves as a Rosetta Stone for the team.**

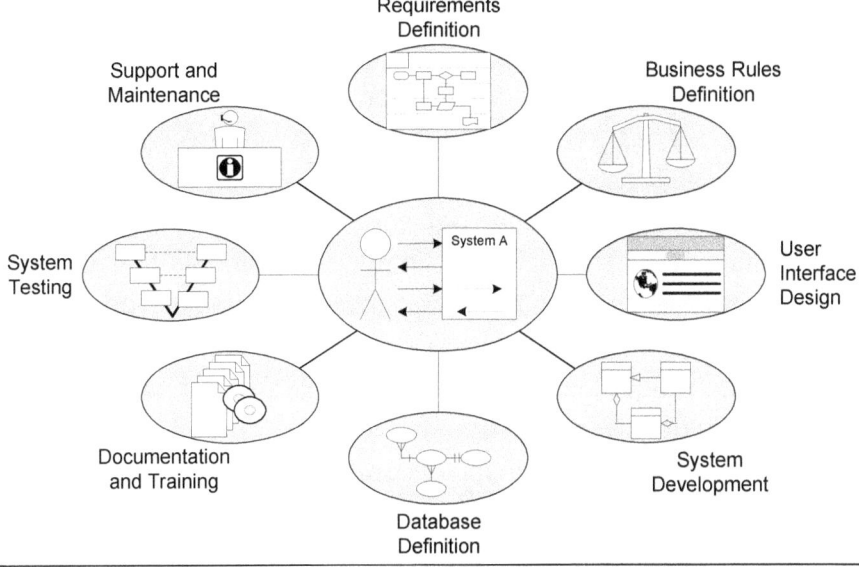

Figure 2: Why Use Cases?

Each of these perspectives can draw from the use cases to better understand the activities necessary to support the user requirements (Bittner and Spence 2003, Armour and Miller 2001).

- The customer can make better decisions about how the system will behave.
- The customer gains confidence that the desired capabilities will be developed.
- Managers can better anticipate the work required to complete the project.
- Analysts can drive out clear functional requirements, business rules, and specifications.
- Usability designers can better understand the users' goals, the flow of the conversations, and their implications in the design of the user interface.
- Architects can make more educated trade-off decisions in system architecture definition activities.

- Developers are able to derive potential classes and methods needed to implement the functional capabilities.
- System integrators can better understand the necessary interactions of multiple computer systems required to achieve the users' goals.
- Technical writers and user education specialists can better understand the way the computer system will behave, and what users seek to achieve so that they can describe it.
- Quality Assurance Analysts can better understand what the system needs to do so they can write better test cases to verify fitness for purpose.
- Support and maintenance people can better understand the support needs of the system.

Inter-functional communication difficulties are common among the major disciplines involved in software projects. Your software teams might run into difficulties because the requirements were not captured well enough. You may also find that the requirements don't help designers and developers build systems that meet them. You can avoid these problems with use case analysis and the unified understanding of the system they provide. Because use cases better bridge this gap, you will build better systems with improved project performance.

> *Because use cases better bridge the gap between requirements and development, you will build better systems with improved project performance.*

To illustrate this point further, review Figure 3. It shows an artifact map that illustrates how use cases inform other parts of the design and development efforts throughout the course of the project. Notice how all of the disciplines derive artifacts from the solution definition workshops that produce use cases.

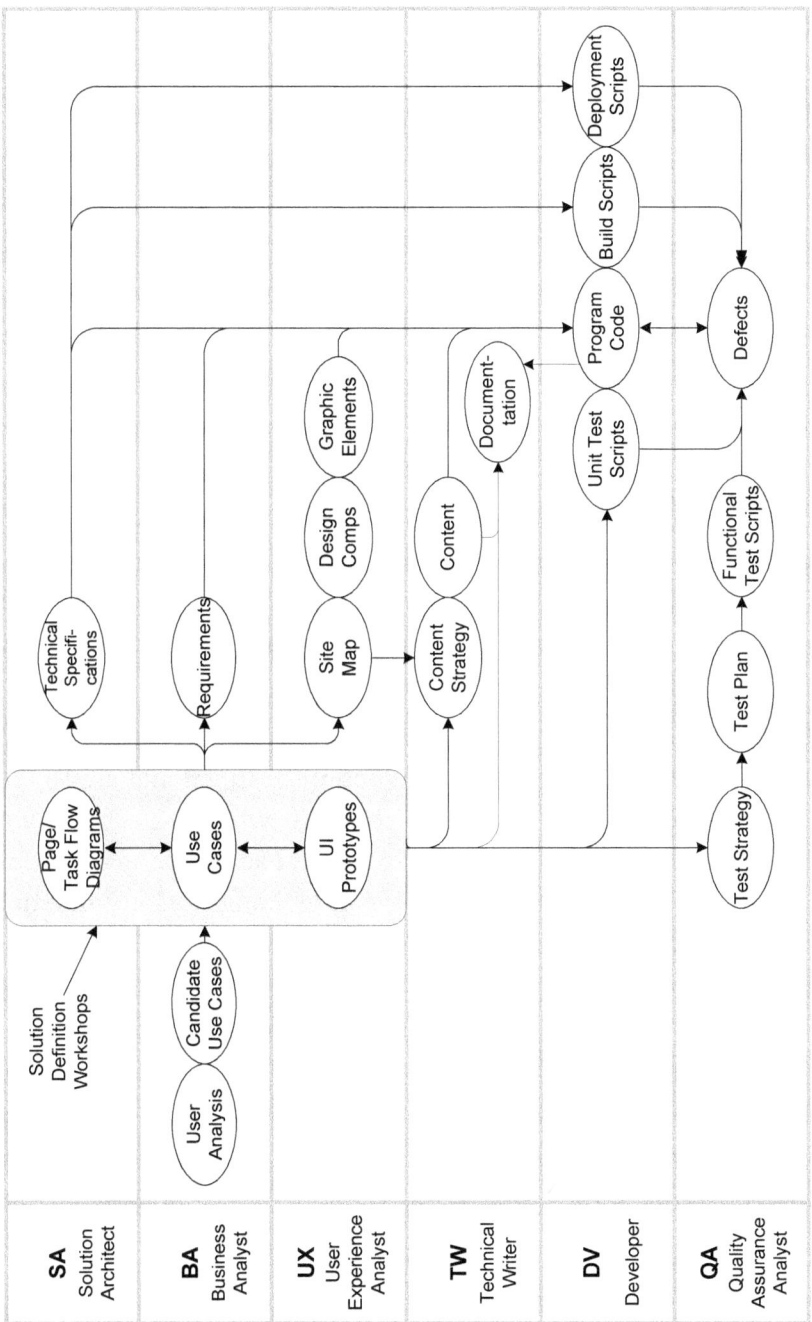

Figure 3: Basic Artifact Map

Readability

Anyone can pick up a use case, read it, and understand what it says. You don't need any special technical or disciplinary expertise to comprehend a use case. This is a very important consideration for business owners who are not experienced in system definition and development.

Scalability

You can employ use cases on projects of any size. For smaller projects, you might prefer the less formal forms of use cases.

Estimating

Use cases and their steps provide you with an excellent base from which to estimate site or system development activities. These include page layouts, design concepts, programs, and integrations.

Scope Management

Use cases provide an excellent definition of functional scope. You can very easily identify scope changes when you base your development estimates on the build-out of use cases. Any additions to the agreed list of use cases are scope changes as are deletions and modifications. Further, these scope changes are easily demonstrated to your customer.

Measuring Progress

When the scope of development is defined as a list of use cases, your whole team and your customers can see for themselves how much progress they have made toward the completion of the project.

Anticipating Failure

You consider possible reasons for failure when you analyze possible exceptions in your use cases. This helps your team to anticipate failure conditions better, to handle those situations more gracefully, or to avoid them all together. The result is better software.

Maintainability

Computer systems last longer than you might anticipate. An up-to-date library of use cases improves your ability to maintain them because you operate from a well-documented foundation of system operations. You can use the library over and over again on all of the projects that follow the system's initial release, updating them each time to reflect the enhancements implemented in the latest project.

Forms of Use Cases

The simplest form of use case you can write is a paragraph of text that describes the conversation the actor has with the system. However, you can use other forms such as diagrams, flow charts, or structured text as needed to help your team establish a common understanding of the system behavior.

Some people get quite rigid in their interpretation of what use cases should look like, what they contain, or how they are documented. Don't worry about that. The purpose of use case analysis is to define system behavior. Form is secondary to that. Do what you need to achieve a common

> *The purpose of use case analysis is to define system behavior. Form is secondary to that.*

understanding. Tailor the formats, headings, and diagrams to address the needs of your team.

Please note that I portray a variety of different systems and scenarios in the examples that follow because I want you to see how they can be applied to different situations. Later in this book, when I discuss how use case analysis fits into the phases of a project, I will employ a single-threaded example so you can get a better feel for the progressive elaboration that takes place through the course of a project.

Use Case Narratives

The following paragraph provides an example of the narrative form of use case to describe the withdrawal of cash from an Automated Teller Machine (ATM).

> The user inserts a debit card into the ATM. The ATM validates the card and requests the user's Personal Identification Number (PIN). The user enters the PIN. The ATM validates the PIN and retrieves the user's identity and account information. The ATM then presents the main menu to the user. The user elects to make a withdrawal from the checking account. The ATM asks the user how much money is requested. The user enters the amount desired. The ATM checks to ensure that the user's balance will cover the request, calculates processing fees, and asks the user for confirmation. The user confirms the request and fees. The ATM delivers the cash to the user and asks if the user wants another transaction. The user specifies that no other transaction is necessary. The ATM ejects the user's debit card and prints the receipt. The user takes the debit card and receipt.

This format feels most comfortable for business users who are not familiar with system development. You can help them get started in the process with this form and add more structure later. You can also write brief narratives to get an early, overall feel for the desired capabilities of an envisioned computer system. Add them to your Project Brief or Charter document to provide a very clear definition of functional scope.

Use Case Diagrams

You can also create Use Case Diagrams to describe the use case model. They are a particular kind of diagram defined in the Unified Modeling Language (UML) standard as shown in Figure 4 (Jacobson, Booch and Rumbaugh 1999).

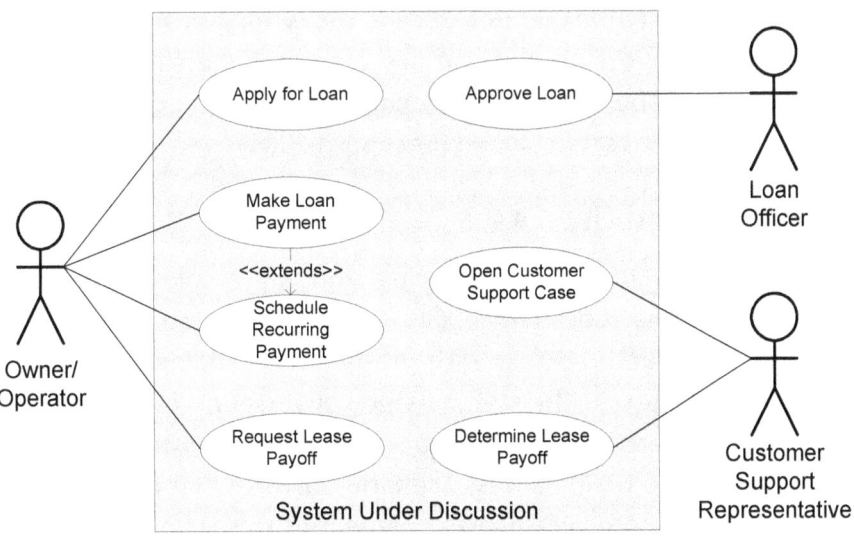

Figure 4: Use Case Diagrams

Use Case Diagrams can be very useful in the early stages of the project to define the system boundaries (Leffingwell and Widrig 2003). You can use them to visualize project scope. They show which use cases will be within the system boundaries and which will not. They can also help to identify relationships among the use cases. For instance, one use case may include another use case, or an alternate path documented in a separate use case may be said to extend the first use case.

As mentioned earlier, your team needs to decide how it implements use cases. Some teams spend a great deal of time getting the use case diagrams correct up front while other teams dispense with use case diagrams in favor of use case narratives or fully dressed use cases. My advice is to be practical. If it adds value for the team, it's great. Don't do it just to conform to some pedantic notion of right or wrong.

The use case diagrams do not replace the writing of the use cases. They only show the goal of the use case, not the substance. The significance of the use case is found in its text. The diagrams just help you find the text you need (Cockburn 2001).

Fully Dressed Use Cases

The most detailed form of use case you will employ is the fully dressed use case (Cockburn 2001). Fully dressed use cases follow the form of structured text with pre-defined headings. Central to the structured text is the step-by-step conversation between actor and system to accomplish the goal. The most straightforward path to success is the Main Success Scenario, or MSS. You or your team might prefer a more light-hearted name for this such as Happy Path or Sunny Day Scenario. The next figure illustrates an example of a fully dressed use case to make an airline reservation.

What are Use Cases?

Use Case Name:	Make a reservation
Primary Actor:	Web User (the user)
Scope:	Airline e-Commerce Website (the website)
Context of Use:	Actor is at home on personal computer or in office environment.

Precondition(s): The user has successfully accessed the airline's website home page.
Successful Post Condition: The user has successfully reserved a seat on a flight.
Minimal Guarantee(s): The website returns a meaningful error message.
Trigger: None

Main Success Scenario (MSS):
1. The user enters a trip request and elects to shop for a flight.
2. The website calls the GDS web service with the trip request information.
3. The GDS searches for matching flights and returns the best matches.
4. The website displays the matching flights anf flight information to the user.
5. The user selects the desired flight and elects to continue.
6. The website requests the passenger information.
7. The user enters the passenger information and elects to make a reservation without purchasing a ticket.
8. The website makes the reservation in the GDS and displays a confirmation.
9. The user leaves the website.

Exceptions:
2a. *There are no available matching flights.*
   2a1. The website advises the user to shop again as no flights are available.
   2a2. The user re-enters a trip request (return to MSS step 1).

Figure 5: Fully Dressed Use Case Example

A quick review of this example conveys a lot about how to achieve the goal of making a reservation. You can tell where the use case begins and where it ends, what system is affected, what success means, and how it is achieved.

You can follow the basic conversation between the user and the website, numbered in a clear series of steps. Notice how the request-and-response conversation moves forward in the Main Success Scenario. You can always tell who the actor is in a given step because the sentences are written in active voice with a subject-verb-object pattern.

A lot happens in these nine steps. Among other things, the website makes two handoffs to achieve the goal. First, "GDS" refers to a

Global Distribution System, a type of system common in the travel industry to record available flights. In this case, the website becomes a secondary actor when it requests information from the GDS. Secondly, the website makes the reservation in the GDS, which in turn updates its database of record for airline reservations.

> *You can always tell who the actor is in a given step because the sentences are written in active voice with a subject-verb-object pattern.*

Notice also that key terms require further definition: Trip Request Information, Flight Information, and Passenger Information. You will drive out additional precision for these terms when you elaborate these use cases later on. Your team might even go so far as to define the method call signatures or interface file formats as part of that elaboration.

Also note that the use case doesn't try to specify the specific user interface. You should define the intent of the actor, not the specific controls that might be invoked by the user such as text boxes, drop-down lists, or command buttons. The user experience analysts will design these when they develop user interface prototypes to accompany the use case.

Finally, note the exception at the bottom. It identifies something that could go wrong and defines what the website does about it. It's normal to have several exceptions in a use case. You define them in more detail later on as well.

Your team will also want to define some alternate paths through the conversation. For instance what if the user wants to pay for a ticket in addition to making a reservation?

You will employ the fully dressed form more often than the others. Help your team to tailor the headings to suit their purposes. For example, sometimes the narrative is included also. You should always include the use case name, the actor, the Main Success Scenario, the alternate paths, and the exceptions.

Other Variations

You may want to supplement some use cases with UML Sequence Diagrams. Sequence diagrams show the multi-layered nature of the conversations necessary to fulfil the goals of complex use cases, as shown in figure 6.

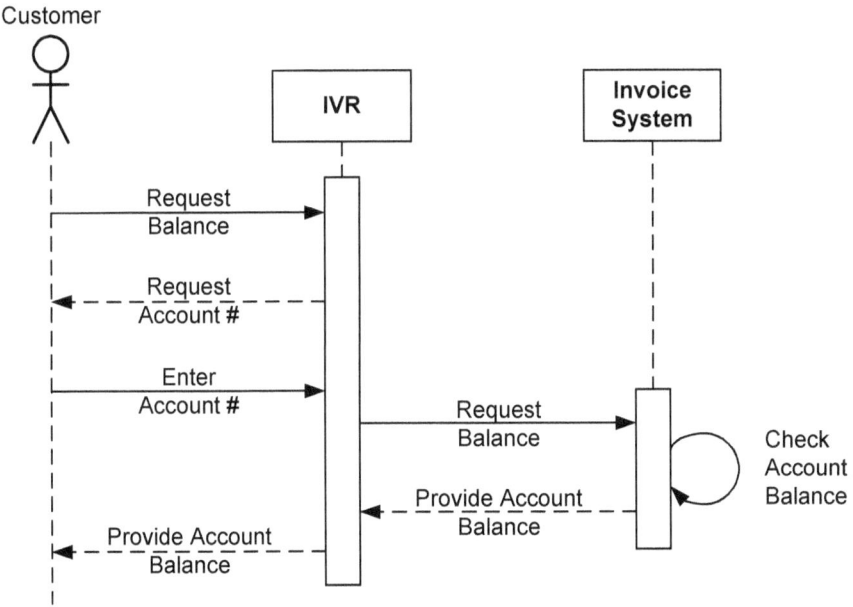

Figure 6: UML Sequence Diagram

Your team might also supplement use cases with flow charts (as shown in Figure 7), low-fidelity prototypes (such as whiteboard screen flows or storyboards), workflow diagrams (with swim-lanes) or UML Activity Diagrams (with or without swim lanes) to describe the flow of events.

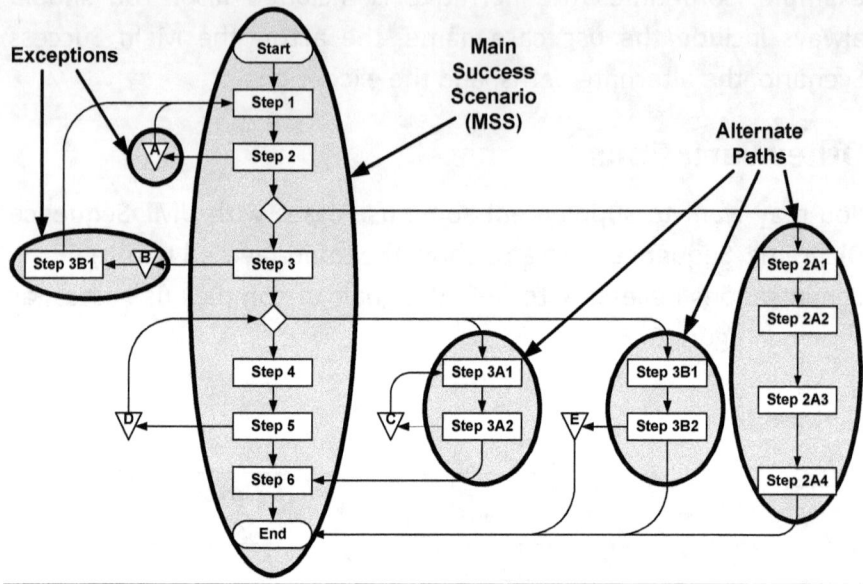

Figure 7: Flow Chart

Remember that you employ use cases to define system behavior. That's what matters. The specific form employed is secondary to that very important objective. You and your team should employ the use case forms that maximize mutual understanding and consensus.

Use Cases Compared to Business Process Modeling

Business Process Modeling documents business processes and seeks to improve the efficiency and effectiveness of those processes through analysis. These process models often portray processes as workflows with swim lanes to identify the roles involved and process steps involved for each role.

Business process models can describe manual processes, automated processes, or some combination of both. Sometimes, you first document an "as-is" process and then transform it through analysis into a more automated, "to-be" process.

People often employ Business Process Modeling in the implementation of Commercial, Off-The-Shelf (COTS) software. For example, you might define 50 or more business processes for the implementation of an Enterprise Resource Planning (ERP) system. The number of business functions affected and the number of modules being implemented will drive this number.

An example of a manual workflow associated with requisitioning is provided in Figure 8.

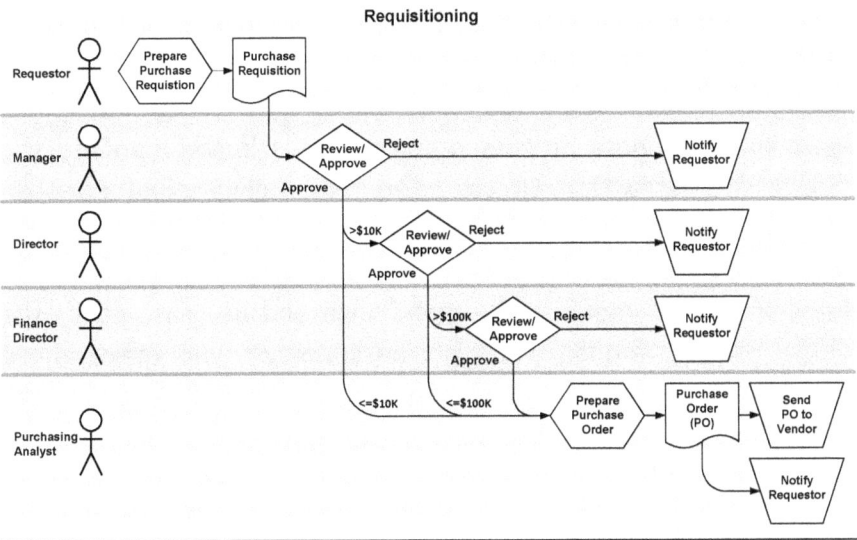

Figure 8: Workflow Model Example

Business process models reflect an organizational view of the process and the interactions among people, roles, or departments in fulfillment of that process. You might say they reflect the "voice of the organization." Use cases, on the other hand, tend to reflect the "voice of the user." Use cases make the business process real and concrete by bridging the gap from organization-oriented processes to user-centric system design, especially when combined with user interface prototypes, wireframes, page flow diagrams, etc.

> **Business processes reflect the voice of the organization while use cases reflect the voice of the user.**

In this example, you might identify several use cases to automate steps in this process as part of a project. For example, you might have a use case called "Review a purchase requisition" with the Manager, Director, and the Finance Director as alternate primary actors. Approving the purchase requisition would be the Main Success Scenario, rejecting it would be an alternate path, and putting ceilings on the amounts that could be authorized would be business rules.

Typically, a business process or workflow is larger than the use cases associated with it. However, you won't always find a one-to-one correspondence between a step in the process and a use case. You might have one use case cover several steps in the process. In detailed process models you might have one use case cover the whole process model. Your business analyst will make those determinations. You should prepare project documentation to map how use cases relate to business processes when business process models are used.

Use case analysis forces you to think about the business process at a deeper level. It covers the interface between the user and the system that supports the process. It also encourages you to think about the management of "state" in each instance of the business process. That is, how will the system maintain awareness and retention of the business data and session management information that flows through the steps in the process?

Sequentially, you would normally complete the business process modeling across several processes before you start use case analysis. This is because business process models drive use case development. In projects where business process modeling occurs, it is not unusual for that to be part of a separate project. You would then perform use case analysis in a follow-on project to automate the processes.

Use Cases Compared to User Stories

If you are an enthusiast of agile and iterative development (Larman 2004), you will recognize that use cases seem similar to the concept of user stories. Indeed, use cases are much akin to the "coarse-grained stories" or "epics" referred to in many agile flavors (Cohn 2004). However, they add value to the concept of stories. They provide additional structure and context to the conversation which, in turn, improves the usefulness of that information for your downstream design, development, and quality assurance activities (Wiegers and Beatty 2013).

Large, complex software projects often require you to change several different systems to support or automate a business process. You need to orchestrate the passage of information back and forth through the systems. Use cases provide you with a more rigorous method than user stories for analyzing the systems integrations to support a user's goal in this complex situation (Leffingwell 2011).

Use cases and user stories both span the layers of a modern computer system as illustrated in Figure 9. The goal of each use case entails interaction of the presentation layer with the user, the execution of business logic, and the storage and retrieval of data. These "vertical slices" represent round-trip transactions that advance the conversation described in the use case. You might require several such round-trip transactions to achieve the user's goal in a single use case. However, you will generally cover only one transaction in a user story.

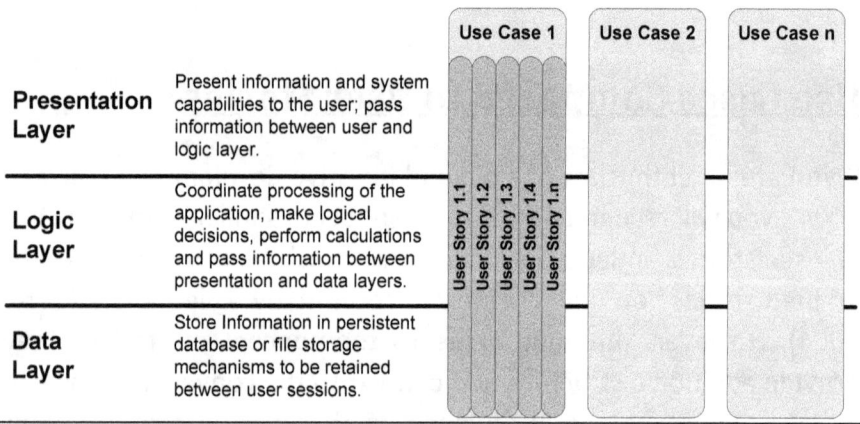

Figure 9: Vertical Slices

Use cases tend to be larger than user stories. User stories are intentionally kept small (Cohn 2004). Well-defined user stories are thin vertical slices that represent the smallest pieces of capability that can be specified, developed, deployed, tested, and remediated. For example, you might define several user stories to fulfill a use case, "Review a purchase requisition," such as:

What are Use Cases?

- As a manager, I can bring up a purchase requisition online, so I can review it.
- As a manager, I can approve a purchase requisition.
- As a manager, I can reject a purchase requisition.
- As a manager, I can annotate my review with comments, so the requestor can see them.
- As a manager, I only approve requisitions for $10K or less.

One way you can define user stories is to set up a story for each possible scenario through the related use case (Cohn 2004). This aligns your user stories and test cases well because test cases also tend to reflect the different scenarios through the use case.

Another possibility is to define user stories for the steps, alternate paths and exceptions defined in the fully dressed use case. Figure 10 provides a visual example of how you can define user stories this way.

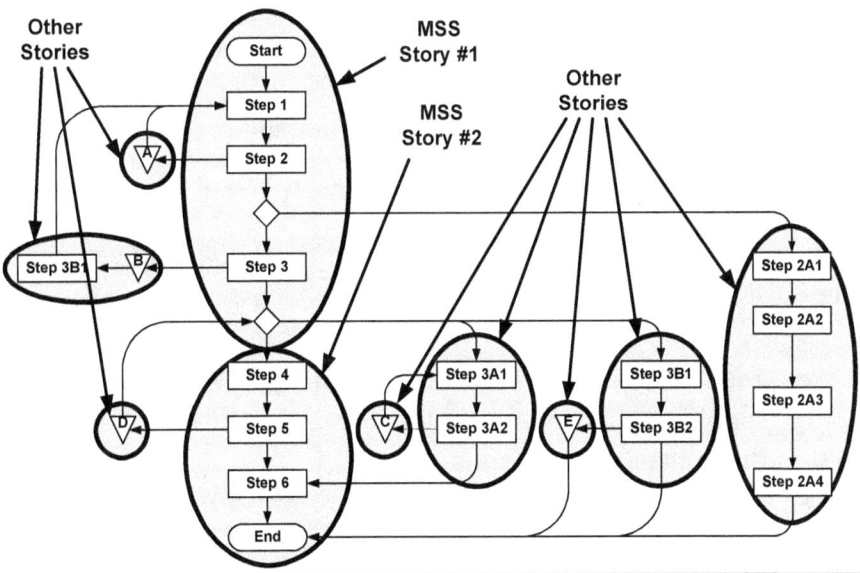

Figure 10: Use Cases into Stories

Both of these approaches allow you to develop the main success scenarios (MSS) of your use cases in early sprints and finish the remaining stories later. This often provides the most business value to the user soonest (Schwaber and Beedle 2002). You can then develop the alternate paths and exceptions in subsequent sprints as prioritized by the product owner.

Another difference between use cases and user stories is the length of time they are maintained. You maintain the library of use cases from the initial development throughout the useful life of the application. However, you only care about user stories for the iteration in which you develop them (Cohn 2004).

Agile advocates might ask, "Why bother with use cases?" Some say that use cases are too much documentation (the "Big-Design-Up-Front" (BDUF) problem that agile seeks to avoid). They may also say that use cases are too much like a contract between the customer and the development team as scorned in the Agile Manifesto. They might argue that they can just define user stories and therefore move into development more quickly. After all, a story is a promise for a future conversation between the developers and the customer.

As you will see in subsequent sections of this book, you can avoid the BDUF problem as follows:

- Restrict the use case definition to the Happy Paths supplemented by a brief list of possible alternate paths in the initial definition sprint.
- Decompose the use case into a set of user stories to build the capability incrementally.

- Elaborate the rest of the use case in the sprint just prior to your planned development of the use case.

I think there are good reasons to exert a bit more extra effort to define the initial use cases first:

- The disciplined and structured approach inherent in use case analysis drives out a more thorough definition of the processes through which and around which the users accomplish their goals.
- Because use cases tend to be larger in scope, combined with the process-oriented analysis, the likelihood of excluding necessary user capabilities is reduced. Too often, projects that "go straight to stories" miss user requirements up front and encounter them downstream after the cost and schedule expectations have already been set with the stakeholder community. Overruns ensue.
- The additional structure of use case analysis over stories provides more complete and useful information to downstream design, development, and quality assurance disciplines and activities.
- The more structured analysis also better facilitates geographically dispersed teams.

Another worthwhile topic is "User Story Mapping." Its steps are:

1. Identify the users of the system you are about to build.
2. List the things they want to do on the system. Those become user stories.
3. Decompose the stories down to a level at which they can be estimated, developed, deployed, and tested.
4. Prioritize the order in which to develop those stories.

This process is very similar to use case discovery. However, it also lacks the deeper comprehension gained in the documentation of the steps for each story.

> *Every team I have worked with that employed use cases has felt they added a lot of value, even those teams that had jumped straight to user stories in prior projects.*

You and your team will need to decide if the added effort of documenting fully dressed use cases is worth the effort. Every team I have worked with that employed use cases has felt they added a lot of value, even those teams that had jumped straight to user stories in prior projects.

When to Employ Use Cases

Your decision to employ use case analysis for a project deserves some reflection. Ideally, your team should decide after objective consideration of several factors.

Development Methodology Standards

Many of the best development organizations have repeatable processes defined as methodology standards for development teams. In some cases, you must follow these practices without much discretion. As a team leader, you should inquire as to whether or not such standards apply. Interestingly, many of these mature organizations include use case analysis as part of their standard approach.

Customer Needs and Desires

Your customer may want to employ use cases because they have used them before or they have seen how they can contribute to a project's success. More often, you will be the advocate recommending them to the customer based on your perception of the customer's needs.

Team Preferences

If your team members have employed use cases before and have become comfortable with them, chances are good that they will want to employ them again. Usually, two or more positive experiences will influence a new team to give them a try.

Complexity of Development Effort

The larger and more complex a project is, the greater need there will be to manage lots of requirements and related information (Leffingwell 2011). Similarly, the more people there are involved in a project, the more differences of opinion there will be about how the job should be done. Use cases can help you on both counts. They provide handy reference points around which to organize requirements. It is quite common for use case analysis to help identify functional requirements. They also serve as excellent forums to discuss and reconcile different opinions about business and system processes. In general, you will find that the more complex the project is, the more advisable use case analysis is.

> *It is quite common for use case analysis to help identify functional requirements.*

Conversational Nature

The conversational nature of use cases make them a natural fit for you to define how a computer system interacts with its users to support their needs. If you are building a system that does not engage in two-way conversations with its users, you will not benefit much from use case analysis.

Project Type

Use cases are better suited to some types of projects than others. Figure 11 shows you if use cases are appropriate for several different types of projects.

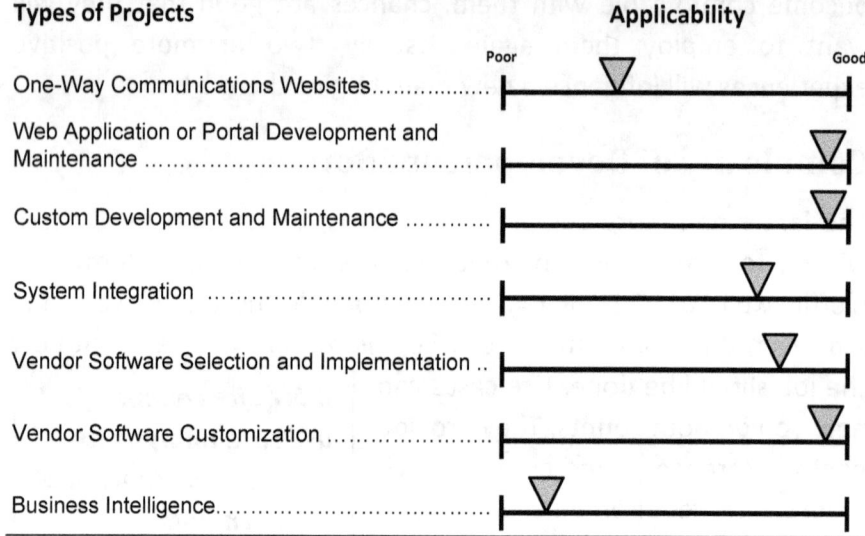

Figure 11: Use Case Applicability

One-way Communications Websites

You might think that use cases would be a good tool for websites in general, but one-way communications websites, the primary function of which is to convey information to the reader, don't have much conversational interaction. This type of website would include most marketing-oriented presentations of company capabilities. There might be a lot of information conveyed to the user, but there is little back-and-forth conversation with the user. Visual design, analytics, and information architecture can be important considerations for these websites, but use cases won't add much value to the thought process.

Web Application or Portal Development and Maintenance

Web applications, which provide significant system capability through the web browser, are ideal candidates for use case analysis. This type of project includes most e-commerce websites, websites that provide online registration, or portals that launch or provide access to other applications. These websites engage in a variety of conversations with their users and that is what makes them ideally suited for use case analysis. You will find use case analysis equally applicable to the initial development of these websites and to their maintenance and enhancement over time.

Custom Development and Maintenance

Closely related to web application development, custom software development refers to applications that you write from scratch to support a business function. These applications tend to converse with users and are therefore good fits with use case analysis.

It is interesting to note that custom software development does not take place as much as it used to. More of the programming that takes place today deals with the customization and integration of vendor software packages.

System Integration

Use case analysis will help you with systems integration work because actors can be computer systems. The old-style mechanisms for systems integration, such as overnight passes of data files, might not be a good fit, but more modern, near-real-time messaging can be.

Vendor Software Selection and Implementation

This is an interesting area for use case analysis. Most people think about business processes when they consider vendor COTS packages, but often overlook the next level of detail in the use cases. As mentioned earlier, business processes reflect the voice of the organization, but use cases reflect the voice of the user.

Even in the selection process, a list of actors and their goals can provide you with an excellent set of objectives and requirements to use in selecting a prospective package. At this high level, they define what the system needs to be able to do without providing the details of how those needs are fulfilled.

You can define those details later in the implementation stage with use cases that are defined in light of the unique capabilities, limitations, peculiarities, and customization potentials of the vendor package to be implemented.

For example, most all accounting systems have the capability to allow an Accounts Receivable accountant to generate an invoice (actor and goal), but each accounting system will have its own unique way of generating an invoice (the details of that use case).

Vendor Software Customization

Use case analysis will be very useful to you in the implementation and customization of a vendor COTS package. The conversations you document will need to be designed with the capabilities and styles of the COTS package in mind. Consider it a technology-constrained approach. Given that, you will want to add a person to your team who knows the vendor package well.

Business Intelligence

Business Intelligence includes data warehousing, reporting, and systems integrations to support the data warehouses. I don't think you will find that data warehousing is a very good fit for use cases. This is because the design of a data warehouse is driven by data definitions, not user conversations. Even the system integration

Extract, Transform, and Load (ETL) programs tend to be one-way data sharing.

Other Considerations

This first section should have given you a good idea of what use cases are and how they can add value to your software projects. Their ability to serve as a Rosetta Stone for your team is central to their value. Nonetheless, use cases do not represent a panacea for all of the problems that can arise in software projects.

You should not rely solely on use cases to define the system requirements. You be will wise to also build a register of functional requirements in parallel with your solution definition workshops. Use cases will flush out requirements and round out the understanding of the functional needs. As your team discusses a use case, you should make notes of requirements to ensure the thoughts are not lost.

You should still interview Subject Matter Experts (SMEs) even though you write use cases. Interviews of stakeholders and experts provide significant insights into the needs and rationale behind a project well beyond a strict list of user requirements. In your interviews, you will learn a lot of information that will help you be successful in the project, such as politics surrounding the project, alternatives that were considered, previous attempts that failed, etc.

Finally, you should supplement your use case analysis with other documents to define non-functional requirements such as security, network performance, data models, etc. Documents such as

Technical Specifications will guide the team in the internal design and development considerations.

Finally, you should develop testing and quality assurance strategies to define how and when the software will be tested. Test strategies are then often followed up by test plans that define the specific tests to be conducted and when.

USE CASE ROLES AND RESPONSIBILITIES

Before discussing how to define use cases, it will be worthwhile to consider the roles of the people on the team who should be involved in the use case development process.

What is a Role?

A role is a part that is played in a project by one or more persons. Each role has responsibilities to be fulfilled by the person(s) associated with it. A person can fill more than one role on the project. For example, Nandini, who is a multi-faceted individual, could act as both the business analyst and the quality assurance analyst for a project. Conversely, more than one person can act in a given role. For example, Quang, Beth, and John might all be developers on a project.

Roles are often confused with positions. Your position is the job you hold in your organization. It is a classification of your job by the Human Resources function. Your title is associated with your position. One person fills one position. Your department can have more than one instance of a position authorized and staffed, but one person has only one position.

A role matrix can help sort this out if your team has difficulty with the concepts, as illustrated in Figure 12.

	Julie	George	Quang	Jenny	Beth	John	Nandini
Product Owner	X						
Project Manager		X					
Solution Architect			X				
Business Analyst		X					X
User Experience Analyst				X			
Developer			X		X	X	
Quality Assurance Analyst	X						X

Figure 12: Role Matrix

Some roles are naturally shared across several people while other roles are more often combined, such as the business analyst and QA analyst roles.

In practice, the determination of which parts people will play is driven by the size of your team, the nature of the work that needs to be done, the versatility and experience of your team members, and your team members' preferences.

The Roles in Use Case Analysis

Generally, the larger the project, the more specialization you will find in the role definitions. For example, a content-rich project might require specialized roles for content strategists, copy writers, and editors. A smaller project, however, might employ a more generalized role such as a Technical Writer to do many of the same things. Similarly, larger development efforts sometimes see developers specialized into presentation-layer, middleware, persistence-layer developers, and systems-integration specialists.

The list of roles discussed here typifies a medium-sized project.

Product Owner (PO)

The Product Owner represents the business owners for the project and is often filled by someone with a position title of Product Manager. The responsibilities of this role revolve around the definition of business requirements and the coordination of efforts in the business community to support the project. This includes stakeholder identification, communication and coordination, business issue research and resolution, business rule definition, and the development of functional test scripts. Some teams assume that the individual serving as the Product Owner is the only voice required. More realistically, the Product Owner represents several stakeholders and must work to ensure that all of their perspectives are aligned for the project.

> *The Product Owner represents several stakeholders and must work to ensure that all of their perspectives are aligned for the project.*

Project Manager (PM)

The Project Manager coordinates efforts of business and technical team members to accomplish the project goals and objectives. This includes writing use cases, scheduling workshops, arranging facilities, and managing the costs and schedules.

Business Analyst (BA)

The Business Analyst is responsible for definition of the business requirements with the detail and precision needed for the developers. Use case analysis is most often associated with the business analyst discipline. The Business Analyst should lead the use case analysis sessions. They also participate in the definition of business rules and more detailed requirements, and in the resolution of business issues for the project.

Solution Architect (SA)

The Solution Architect defines the architecture for the system and works throughout the project with team members to preserve the conceptual integrity of the system as the project progresses. The Solution Architect participates in use case analysis discussions to:

- Understand the business requirements and to ensure that the envisioned solution addresses those requirements.
- Help the team analyze trade-offs between design and cost.
- Offer alternative design approaches that could be accomplished more inexpensively.

In addition, the Solution Architect has the primary responsibility to capture non-functional requirements that often emerge in use case discussions.

Developer (DV)

The Developer takes part in use case analysis and associated design activities to represent the development perspective in those discussions, to elicit requirements from the business community, and to capture non-functional requirements as they arise. If developers are not yet on board when the use case analysis takes place, the Solution Architect represents the developer's perspective.

User Experience Analyst (UX)

The User Experience Analyst, also referred to as Usability Analyst, Information Architect, User Interface (UI) Designer or Interaction Designer, provides expertise in the design of a pleasant, intuitive and positive user experience for the system. The UX analyst prepares wireframes or UI Prototypes for the team to review while discussing a use case.

Quality Assurance (QA) Analyst

The Quality Assurance (QA) Analyst participates in the use case analysis process to understand requirements as they emerge, to ensure that requirements are valid and testable, to formulate test strategies and plans, to capture information for test scripts, to track fulfillment of requirements, and to drive the development of the functional test scripts.

Technical Writer (TW)

The Technical Writer provides the user documents, help files, training materials, or desk procedures that often accompany interactive software systems.

Responsibility Matrix

Figure 13 illustrates a sample of the responsibilities for each of these roles in a project that employs use case-driven development.

Use Case Analysis

WBS	Task	PO	PM	BA	SA	UX	DV	QA
1.0	Sprint 0 - Definition							
1.1	Project Initiation							
1.1.1	Project Brief	A	O	C	C			
1.1.2	User Analysis	A	C	O	C			
1.1.3	Use Case Discovery	A	C	O	C			
1.1.4	Staffing Confirmation	A	O					
1.2	Solution Definition							
1.2.1	Initial Requirements Analysis							
1.2.1.1	Define Initial Use Cases	A		O	C	C		C
1.2.1.3	Define Initial Functional Requirements	A		O	C	C		C
1.2.1.3	Define Initial Business Rules	A		O	C	C		C
1.2.2	Initial Design Analysis							
1.2.2.1	Define Initial Site Map	A		C		O		
1.2.2.2	Define Initial Wireframes	A		C	C	O		
1.2.3	Initial Technical Analysis							
1.2.3.1	Write Initial Technical Specifications	A			O		C	C
1.2.3.2	Define Non-Functional Requirements	A			O		C	C
1.2.4	Initial Issue Analysis	O	C	C	C			
1.3	Project Plan Update	A	O	C	C	C	C	C
2.0	Sprint 1 - Development							
2.1	Elaborate Sprint 1 Use Cases	A		O	C	C		C
2.2	Design Mockups for Sprint 1 Use Cases	A				O		
2.3	Build Out Sprint 1 Use Cases	A		C	C		O	
2.4	Write Test Scripts for Sprint 1 Use Cases	A		C	C			O
2.5	Test/Fix Sprint 1 Use Cases	A		C	C	C	C	O
2.6	Elaborate Sprint 2 Use Cases	A		O	C	C	C	C
2.7	Design Mockups for Sprint 2 Use Cases	A				O		
3.0	Sprint 2 - Development							
3.1	Build Out Sprint 2 Use Cases	A		C	C		O	
3.2	Write Test Scripts for Sprint 2 Use Cases	A		C	C			O
3.3	Test/Fix Sprint 2 Use Cases	A		C	C	C	C	O
3.4	Elaborate Sprint 3 Use Cases	A		O	C	C	C	C
3.5	Design Mockups for Sprint 3 Use Cases	A				O		
4.0	Sprint 3 - Development							
4.1	Build Out Sprint 3 Use Cases	A		C	C		O	
4.2	Write Test Scripts for Sprint 3 Use Cases	A		C	C			O
4.3	Test/Fix Sprint 3 Use Cases	A		C	C	C	C	O
5.0	Sprint 4 - Testing & Deployment							
5.1	Full System Integration Testing	A	C	C	C	C	C	O
5.2	User Acceptance Testing (UAT)	A	C	C	C	C	C	O
5.3	Solution Staging	A	C		O			
5.4	Conduct Training	A		O	C	C		
5.5	Launch into Production	A			O		C	C

ORCA: **O** = Own, **R** = Review, **C** = Contribute, **A** = Approve

Figure 13: Responsibility Matrix

Use Case Roles and Responsibilities

The Responsibility Matrix is a useful tool to help you define, communicate, and estimate the responsibilities of each of your team members throughout the project. It has a column for each role and a row for each task or deliverable. The intersecting cells convey the responsibility that the role has for the task. Note that entries are made for the detail-level tasks alone, not the summary levels.

> *The Responsibility Matrix is a useful tool to help you define, communicate, and estimate the responsibilities of each of your team members throughout the project.*

A scan down a column shows all of the tasks with which a role is associated and a bit about the responsibility of that role for that task. A scan across a row shows all of the roles involved in the completion of the task and how each role is involved. This can be useful to help your team members understand their parts in the project.

The entries in the cells of this particular responsibility matrix form the acronym, "ORCA":

- O = Owner – the person who owns, or is accountable for, the work product of a task.
- R = Reviewer – a person who reviews the artifact(s) produced by a task.
- C = Contributor – a person who helps the owner produce the artifacts generated by a task.
- A = Approver – the person who has the authority to accept the work results from a task.

The most common legend is "RACI", which is an acronym for Responsible, Accountable, Consulted, and Informed. Indeed, many people refer to these charts as RACI charts.

This particular responsibility matrix identifies the typical responsibilities for the tasks in the Work Breakdown Structure (WBS) of a quasi-agile project life cycle that supports use case analysis.

Here are some general observations about this responsibility matrix:

- The Product Owner is the approver of the work on the project. This makes sense if you consider that the product owner represents the business ownership of the project's results.
- The Business Analyst is the owner of most of the use case work.
- The Solution Architect owns most of the technical specifications and non-functional requirements work.

Keep this responsibility matrix in mind and refer back to it as we discuss the steps in developing use cases. You will see how each of these roles contributes to the definition of use cases throughout the remainder of this book.

Developing Use Cases

Use case analysis is a process of discovery. As your team moves through each layer of discovery, it learns more about what it needs to accomplish. New awareness emerges that provides you with improved insight into the envisioned system and associated requirements. Because of this learning process, you should consider it healthy to find that you need to revise your initial work products, that your approaches change midstream, or that your assumptions need to be changed.

Typically, you start with an initial list of potential events, actors, and use cases. I call them candidate use cases at this point because you don't know at this early stage which of them will survive the downstream analysis. You revise this list as your team drills down into the details of the individual use cases. You add use cases to the list, delete use cases from the list, combine use cases, split apart use cases, and revise them along the way.

Use cases evolve through the course of a project. Project management lexicon refers to this as "progressive elaboration." The closer you get to completion of the project, the clearer the vision of the end system becomes and the more you know about it.

> *Your vision starts out rather foggy, but as you begin to define the system's characteristics, you start to see things more clearly and precisely.*

Think of it as walking down a path toward a house on a foggy day. At 100 yards, you can barely see that a house sits at the end of the path. However, as you walk closer, you can make out additional details about the house. It's a two-story house. It has three windows across the top and two on either side of the door. You can see that it is a Craftsman-style home with shutters around the windows and so on.

So it is with the design of a system for the customer. Your vision starts out rather foggy, but as you begin to define its characteristics, you start to see things more clearly and precisely.

In software engineering, the concept is characterized by the "Cone of Uncertainty," as illustrated in Figure 14. Barry Boehm first described it (Boehm 1981) and Steve McConnell refined and named it (McConnell 1998 and McConnell 2006).

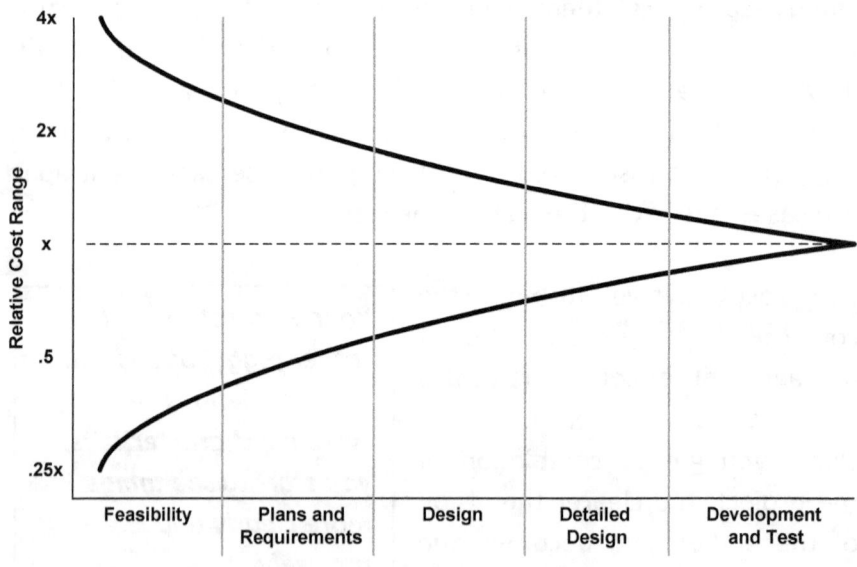

Figure 14: The Cone of Uncertainty

Essentially, the Cone of Uncertainty tells you that estimates you make early in the life cycle of a project might be far off the mark, but that your estimates improve as the project progresses. Of course, this requires that the activities you accomplish in the project are effective at reducing uncertainty.

> *The Cone of Uncertainty illustrates that estimates made early in the life cycle of a project might be far off the mark.*

Use cases are effective at this. You can start use case discovery in the feasibility and discovery stages and then elaborate on them in each of the subsequent stages in the project from requirements through design, development, and testing.

Even a simple list of the presumed users and the associated candidate use cases in the feasibility or discovery stage clarifies the nature of the system and the scope of the project. With practice, they make an excellent basis for estimating the cost and schedule of the project.

As you continue to define initial use cases that are high level and user-oriented conversations, you further clarify the intended operation of the system under development. You start with the initial list of candidates identified in discovery.

Elaboration of those use cases in the design stages provides you with even more specific guidance as they evolve to reflect more system-oriented process definitions. You can drill down into more detail as needed to specify programming logic, input/output definitions, methods signatures, and even pseudo code in some cases.

The level of detail to which you drive use cases depends on the inclinations and needs of your development teams. Some teams are inclined to develop detailed specifications. You will find this especially true when teams are dispersed around the globe. Other teams, especially agile project teams, may choose to curtail the specifications activity at a higher level and instead rely on frequent customer conversations to gain the deeper levels of comprehension needed to provide a solution that fits the needs of the customer.

Precision Levels

The term, Precision Level, reflects the discovery layers in the progressive elaboration process. It is drawn from the concept of precision stages in Alistair Cockburn's excellent book, "Writing Effective Use Cases" (Cockburn 2001). I recommend six precision levels.

- Precision Level 0 (Discovery) – This base level of precision is essentially a list of the candidate use cases. Included are their primary actors, brief narratives, and definitions of the context in which the use case occurs (the context of use).
- Precision Level 1 (Scope) – Precision level 1 adds to each use case: The system under discussion (the scope), stakeholders, pre-conditions, successful post-conditions, minimal guarantees, and triggers.
- Precision Level 2 (Main Success Scenario) – Also known as the Happy Path or the Sunny-Day Scenario, precision level 2 defines the steps in the Main Success Scenario.
- Precision Level 3 (Alternate Paths) – Alternate paths are added to the use case to define branches off the Main Success Scenario to handle varying conditions, user decisions, and logic changes.

Developing Use Cases

- Precision Level 4 (Exceptions) – Exceptions define the anticipated problems and error conditions as well as the system's responses to those exceptions.
- Precision Level 5 (Supplemental Information) – Level 5 provides additional information at any stage in the process to capture such things as user-interface mock-ups, open issues, test script considerations, technology considerations, security considerations, validation plans, and general notes from the discussion.

Your analysis for one precision level might lead you to revise earlier precision levels. You may also find yourself merging, splitting, and revising use cases. This is normal and positive. It is a sign that your team is thinking things through with more precision. Consider that each increase in precision level reflects an increase in the team's collective understanding of the required capabilities and a decrease in the uncertainty associated with those capabilities, as illustrated in Figure 15.

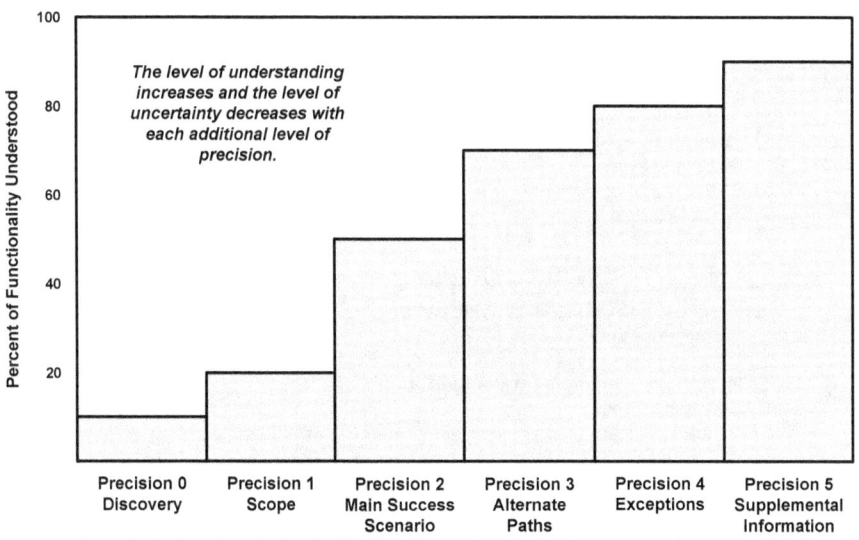

Figure 15: Precision Levels

These levels of progressively more detail and deeper analysis manifest themselves in the major sections of a fully dressed use case, as illustrated in Figure 16.

PRECISION LEVEL 0 – Discovery
Use Case Name: Review a Purchase Requisition (PR)
Primary Actor: Manager (a second or third-level supervisor in the organization)
Narrative: A line manager receives an email noting that a PR is waiting for review. The line manager clicks on the link, reviews the PR, and chooses to approve it, reject it, or send it back for clarification with remarks for the requestor.
Context of Use: Typical office environment on a laptop during normal working hours.

PRECISION LEVEL 1 – Scope
Scope: Purchasing System
Stakeholders and Interests:
1. The requestor would like to get the requisition approved.
2. The Procurement organization wants to ensure that procedures are followed.
Precondition: The manager has opened an email requesting the review of a PR.
Success Post Condition: The PR has been reviewed and approved.
Minimal Guarantees: The Purchasing System provides a meaningful error message.
Trigger: Arrival of an email from the Purchasing System.]

PRECISION LEVEL 2 – Main Success Scenario
1. The manager clicks on the link in the email.
2. The Purchasing System presents the PR for review.
3. The Manager reviews the PR and approves it.
4. The Purchasing System updates the PR status and confirms it back to the Manager.
5. The Manager closes the email.

PRECISION LEVEL 3 – Alternate Paths
3a. The Manager reviews the PR and rejects it.
    3a1. The Manager enters notes to the requestor and rejects the PR.
    3a2. Return to step 4.
3b. The Manager reviews the PR and asks for clarification.
    3b1. The Manager enters notes to the requestor and asks for clarification.
    3b2. Return to step 4.

PRECISION LEVEL 4 – Exceptions
2a. The Purchasing System can't find the PR.
    2a1. The Purchasing System provides an error message to the Manager.
    2a2. Return to Step 5.

PRECISION LEVEL 5 – Supplemental Information
Security Considerations:
- The Purchasing System uses the company single-sign-on protocols.

Figure 16: A Fully Dressed Use Case

For a more complete view of the precision levels in a use case, you might want to take a look at the use case template in Appendix A.

You define these six precision levels in a basic three-step process:

1. Use Case Discovery – Your team defines a master list of candidate actors and use cases (precision level 0) in the authorization stage or during project charter development.
2. Initial Use Case Definition – The essential use cases (precision levels 1 and 2) are defined in the requirements stage or the initial sprint.
3. Use Case Elaboration – The use cases are finalized at the level of detail necessary to provide development specifications (precision levels 3 through 5) in the design stages or in the development sprints.

Each of the three steps is next discussed in more detail.

Use Case Discovery

As the name implies, the first thing you need to do is to discover which use cases will be necessary. This discovery process results in a list of actors and the associated candidate use cases to be developed for the project. The list is documented in a Use Case Register which may be kept in a spreadsheet or similar tool.

> *Discovery activities vary from one project team to the next.*

Discovery activities vary from one project team to the next. In its simplest form, you might individually list the use cases you have discerned from a Request for Proposal (RFP) or from interviews with your customers. On the other end of the formality spectrum, you might conduct surveys and focus groups to identify actors and their goals. Your formality decision will be driven by several factors: The nature, size, and complexity of your project, and the desires of your stakeholder community. For example, you would engage in extensive market research to define the use cases for a new mass-market software product.

In large projects, use case discovery can be a complex task that stems from multiple rounds of review and revision. It depends on the size and complexity of the project, the layers of management involved, and the delegation culture of the organization. In those situations, a separate discovery engagement or statement of work (SOW) might be recommended.

In most cases, you should be able to complete your discovery of use cases in one or two meetings with the right people involved. You will want to get representatives from the business owners, the user community, key stakeholders, and the development project team. These meetings are best conducted by a skilled business analyst, but you may need to do this yourself if time or budgets are limited.

The steps in use case discovery include:

1. User Analysis
2. Event Modeling
3. Listing Candidate Use Cases

It will be helpful to focus the discussion and examples for these and subsequent topics on a single sample application. This sample system is a flight-line tool room automation system called the "Automated Tool Tracking System" (ATTS). At the beginning of each shift, aircraft mechanics report into the tool room to get the tools they need for their shift. Sometimes, these are just toolboxes with standard tools for aircraft mechanics. However, sometimes the mechanics need special tools to accomplish the jobs they have been assigned for the day. Some of these tools can be quite sophisticated and expensive. The mechanics check out their tools for the day, work on their assigned jobs, and check the tools back into the tool room at the end of their shifts. It is very important to account for every tool every day to avoid tools getting ingested by the aircraft engines the next time they start up. Our sample system relies on barcode labels and scanners to make the check-out and check-in transactions as quick as possible.

User Analysis

Start your discovery of use cases with an analysis of the users of the envisioned system. With your team, visualize and discuss the different categories of users of the system under discussion. This helps to identify actors. User analysis can also lead to personas and user interface mock-ups.

It will help to begin these discussions with a review of the vision for the system followed by a brainstorming exercise to identify types of users, or user roles. You should write down the ideas on a flipchart or in a document that is projected onto a screen for all to see. Alternately, the members of the team might write down possible user roles on index cards, one card per user or type of user, and then add them to a central pile of cards.

For instance, our tool room application, ATTS, will have users such as Tool Room Operators and Tool Room Supervisors, Remote Tool Room Operators, Mechanics, and Repair Center Supervisors. These may directly serve as actors, or you may need to modify them as you begin to identify their goals for working with the system.

You and your team should also discuss several questions to help identify other actors (Schneider and Winters 2001).

- What actors create, read, update, or delete information in the system?
- What actors need to be notified of changes?
- What actors involve the system in response to external events?

When you have exhausted your ideas, organize the lists or cards to consolidate duplicates and to group together related roles. Then, describe each of the roles in more detail so that all of your team members can visualize these types of users in a meaningful way. Some things to consider are:

- How computer-proficient are these users?
- How often do they use the system?
- What is their physical environment when they use the system?
- What is their emotional state when using the system?

You should try to name actors in ways that describe their responsibilities. This is considered a best practice for use cases. Be careful, however, to avoid organizational position titles or organizational roles as the names of actors (Bittner and Spence 2003). Instead, name their roles as they apply to their usage of the system. For example, Tool Room Operator, Remote Tool Room Operator, and Repair Center Supervisor may all actually fit well within the responsibility set of an actor called "Operator."

> *You should try to name actors in ways that describe their responsibilities.*

Don't forget to discuss actors that are computer systems. In modern computing environments, almost all processes cascade through an ecosystem of integrated applications. As such, these systems often become secondary actors that request data from other systems or push data to other systems. Some examples would include:

- Web Portal
- Integration Hub
- Interactive Voice Response (IVR) System
- Case Management System

Event Modeling

You should also discuss events that might occur related to the system under discussion. The collection of events that affect the state of the system is referred to as the event model. It helps to derive the actors and goals associated with a system as the triggers of the events. Examples of events might include:

- Each night at 1:00 a.m., the airline's mileage plan system checks the FTP site for files from partner airlines to update mileage plan balances.
- A purchase order arrives on the company's FTP site, which triggers the "PO Listener" to update the order entry system.
- 15 days before due date, the online payment system emails the customer with a notice that a payment is almost due.
- Every 30 seconds, the system refreshes the cache with the list of tools checked out from the tool room.

Events often trigger an actor to have a goal. You can think of them as, "An event happens which causes an actor to do something." In many cases, these actors are computer systems.

Listing Candidate Use Cases

After you have completed your user analysis and event modeling sessions, you can then engage in team sessions to identify the candidate use cases. For each actor identified, you and your team should ask yourselves several questions (Schneider and Winters 2001).

- What goals does this user want to accomplish with this system?
- What information is created by this actor as the system is used?
- How is that information maintained?
- What other systems interact with this system?
- How and when is that interaction initiated?

As with any organized brainstorming session, be sure to focus on the collection of ideas without an evaluation of them until all ideas have been exhausted. Evaluating ideas "on the fly" is a sure way to kill the creative expression of brainstorms. There will be ample opportunity to evaluate, consolidate and clean them up afterwards.

> *Evaluating ideas "on the fly" is a sure way to kill the creative expression of brainstorms.*

Once you have discussed all of the actors, hold a similar discussion for each of the events your team has identified. Ask yourselves, "What actors are affected by this event and what are they triggered to do as a result of this event?"

After your team has completed the brainstorm sessions, you should go back and clean up the list of possible use cases. You should discuss each use case listed to be sure everyone understands it. You should ensure that the goal of the use case is important to the actor. This helps to ensure that the system will be usable and that it fosters improved adoption by the user community. This may lead you and your team to delete some of the candidates. You should also look for duplicates in the list and merge them to eliminate duplication. You might also want to restate the names of the use cases to maximize clarity in as few words as possible. It is always easier in the long run to use short titles that are supplemented with other descriptive text than to get verbose in the titles of the use cases.

Other things to consider for use case names include (Armour and Miller 2001):

- The use case name should be short and describe the actor's goal in plain language.
- The use case name should be stated from the actor's perspective.
- The use case name should be an active verb-noun phrase.

For example, "Set up a Remote Tool Room" is a good use case name. "Remote Tool Room Initialization" is not.

You should now have an initial list of candidate use cases. They are candidates because your further analysis will result in several changes to the list as some are expanded, some are combined, some are split apart, and some are deleted. This is all a good sign of progress on the journey of discovery about the functional capabilities of the system under development.

> *Just an hour or two of discovery can greatly improve the conception of the system shared by the producer and customer team.*

It is always amazing to witness how much understanding about the system is gained from these exercises. Just an hour or two can dramatically improve your conception of the system.

Furthermore, if you do this before estimating project costs and schedules, you will be able to estimate much more accurately because you have an improved understanding of what you will be building. Each use case can be estimated based on its perceived complexity, as discussed later in the section, "Estimating with Use Cases."

You will also have a clear statement of functional scope, which can be stated explicitly, finitely, and unambiguously as in, "Project XYZ will build out these use cases …." This sets an excellent foundation for the recognition of scope changes and their management later on in the project. If you are managing a consulting project, clearly identifying scope changes can protect the project's profitability significantly.

Use Case Register

As you conduct your discovery sessions, you should capture the information in some way that will facilitate the maintenance of that information over time. I recommend a listing I call the Use Case Register. Figure 17 provides an example.

Developing Use Cases

UCID	Event	Primary Actor	Goal (Use Case Name)	Brief Narrative	Priority	Target Release
UC01	N/A	Operator	Check out a tool	The tool room operator checks a tool out to a mechanic	Very High	R1
UC02	N/A	Operator	Check in a tool	The tool room operator checks a tool back in	Very High	R1
UC03	N/A	Supervisor	Assign a new badge	The supervisor assigns a new badge to a mechanic, operator, or supervisor	High	R1
UC04	N/A	Operator	Submit Missing Tool report	The operator completes a missing tool report if a tool is lost	High	R1
UC05	N/A	Supervisor	Conduct shift change	The supervisor records a shift change turnover to another supervisor	Medium	R2
UC06	Every 30 Seconds	ATTS	Refresh open tools list	The system refreshes the memory-resident list of tools currently checked out	Medium	R2

Figure 17: Use Case Register

Give each use case a unique identifier (ID) to facilitate cross-references among the set of use cases. The format of this ID number is whatever works for your team. A prefix before the serial number like "UC" quickly identifies this artifact is a use case. Similarly, "R" in front of the serial number designates a requirement, "I" designates an issue, and "BR" designates a Business Rule.

You should also define a few more attributes for each use case to complete the "Precision Level 0" definition of the use cases:

- Event – If a use case is triggered by an event, you should describe that event.
- Primary Actor – Name the primary actor that initiates the use case.
- Alternate Primary Actors – List other actors that might seek the goal stated by the use case name.

- Brief Narrative – Emphasizing "brief," describe what happens in the use case.
- Context of Use – Describe the technical and environmental context of the actor when the use case is initiated. It can also address the actor's intent, motivation, experience, and expertise.

Furthermore, you may add other fields to the register to suit your needs. For instance, you could add columns to reflect:

- Use case category
- Priority relative to other use cases
- The release in which you envision building out the use case
- Status of the use case
- Requirement ID numbers associated with each use case
- Associated Business Rule ID numbers
- Associated Test Case ID numbers
- Labor estimates by skillset, such as BA, QA, or Dev

As a convenience, you can define each use case name or Use case ID number as a hyperlink that points to the section of the use case document or wiki page that details the use case. In addition, you can link the actor entry for each use case to a section or list that provides the user analysis for that actor.

You will also need to choose the tool or storage mechanism in which you record the use case register. Some ideas include:

- A spreadsheet, possibly within a workbook that also contains spreadsheets for requirements, issues, business rules, etc.
- A word processing document
- A wiki site or similar collaboration tool

- A web page
- A relational database
- A table in a requirements management tool
- The backlog feature in an agile management tool

The spreadsheet or word processing document may in turn be kept on a document management system, wiki site, or another collaboration tool to facilitate sharing and to address document management concerns.

Personally, I start with a spreadsheet because I almost always calculate estimates by skillset in the use case register. However, I will use any of the other options if that is the preference of my customer or team.

Initial Use Case Definition

Initial use case definition occurs after the project plan has been approved and the definition stage of the project has begun. The term, definition, encompasses both requirements and initial design. Requirements and design are more intertwined than sequential. They both contribute to the definition of the intended capabilities and operations of the system to be developed for the customer.

Requirements and design are more intertwined than sequential.

In agile projects, you might want to employ a short definition sprint that focuses on creation of the project backlog. Without becoming a "Big Design Up Front," a little time spent focused on the overall landscape of the project before detailed discussions contributes to your success.

The intent of "Initial Use Case Definition" is to do just enough of the use case work up front to capture the overall project topography. At this point, you define Precision Level 2, the Main Success Scenario, for each of the candidate use cases that you listed in the use case discovery process. Recall that the Happy Path defines the principal sequence of events to accomplish the goal of each use case without branches, exceptions, or errors. When you have completed this, you should also list the anticipated alternate paths and exceptions in each use case but not define their steps. This gives you a good handle on the extent of complexity associated with each use case so you can better define the stories required to build it out.

To accomplish this, you should set up a series of workshops to discuss the use cases. You might facilitate these workshops personally or task the business analyst of product owner with this responsibility. As a best practice, you should project the use case through an LCD projector (or similar mechanism) so your whole team can see how the use case takes shape and to make suggestions for completion of the use case. This encourages participation of the attendees and speeds up the buy-in of the results. Wiki pages are a convenient way of doing this, as they automatically track revisions, but word processor documents can serve as easily.

As an alternative to team meetings, you could have one person (most often the business analyst), be responsible for drafting the use cases based on interviews with business representatives. The analyst then reviews them with the affected disciplines (design, development, and QA) and the broader stakeholder community. You might consider this approach more efficient, but its "throw it over the wall" nature comes at the price of reduced collaboration and less agreement. As result, you may incur missed nuances, suboptimal design, missed requirements, and downstream rework.

The First Pass

First pass means that you should define Precision Level 1 (Scope) information for all use cases before moving to Precision Level 2 (Main Success Scenario) for any of them. Once the first pass is complete, you and your team then execute a second pass through all of the use cases, defining the Happy Path for each of them.

> *You should update the use case register throughout this process to reflect any changes made to the list.*

This approach allows you to validate the primary actors, goals and conditions for all use cases more quickly, as well as add or delete use cases, before you drill down into the next level of detail for each. Furthermore, some stakeholders will want to participate in the definition of the first precision level, but then leave the added detail to others. You should update the use case register throughout this process to reflect any changes made to the list.

Previously, you recorded your use case information in the Use Case Register. Starting with this first pass, you and your team begin to define the content of each use case. Your focus changes from a list of use cases to discussion of each use case individually. You and your team will need to decide where and how you will record the use case contents. Some options would include:

- One spreadsheet for each use case.
- One word processing document for each use case.
- A workbook containing a spreadsheet for each use case.
- A requirements management tool.

- A word processing document containing all of the use cases as separate sections in the document.
- A wiki site (or similar collaboration tool) with a page for each use case.

I have listed these options in order of my preference from the least favorable to the most favorable. Spreadsheets are cumbersome for use case content because you work mostly with structured text in a use case. Separate files for each use case can get messy quickly, even if everyone agrees to a shared repository for those files. This applies equally to word processing documents and spreadsheets. Requirements management tools are favored by some but many people end up discarding them for being too rigid or not worth their cost. Writing all use cases in one word processing document can work well if you manage it on a collaboration site. Wiki sites can work well, too. These last two options make it easy to put in hyperlinks for cross references.

To get started on initial use case definition, pick the first use case you want to tackle and transcribe the discovery information (actor, title, events, description, etc.) from the use case register to the use case itself. You can work on the use cases in any order you prefer. When I work with teams, I usually get the basic template in place for each of the use cases to be written and transcribe the information from the use case register ahead of time. It just makes it easier for the team to focus on the analysis and content without the need to fuss with formats.

Figure 18 shows the Discovery Level information for the use case in the running sample, automated tool room application which documents the check-out transaction. The goal in this case is to check out a tool to a mechanic who has just come on shift and is about to go work on an airplane.

Developing Use Cases

PRECISION LEVEL 0 – Discovery
Use Case Name: Check out a tool to a mechanic
Primary Actor: Tool room operator
Narrative: A mechanic comes up to the tool room checkout counter to request a tool for the upcoming shift. The tool room operator retrieves the tool from its storage location and checks it out to the mechanic and records the complete transaction.
Alternate Primary Actors: Tool room supervisor
Context of Use: The operator is standing behind the counter of the flight-line tool room, which is usually caged behind protective screening. It is located in a restricted area, open only to people authorized to access the flight line.

Figure 18: Sample Use Case – Precision Level 0 – Discovery

Most of this information should come from the use case register. Fill in whatever you are missing. Next, you and your team will define the information for Precision Level 1 (Scope). You should have this use case displayed on the wall or screen so everyone can see what you type. That way, you can all discuss the use case as needed while it is being defined and everyone can see what is being recorded. Figure 19 shows the next section of the check-out sample.

PRECISION LEVEL 1 – Scope
Scope: Automated Tool Tracking System (ATTS)
Stakeholders and Interests:

Stakeholder	Interest
Mechanic	Wants the tool, in a hurry!
Tool Room Supervisor	Wants the information recorded accurately and quickly.

Precondition: The mechanic has approached the tool room counter and asked for a tool.
Successful Post Condition: The mechanic has the tool and the system has recorded the transaction accurately.
Minimal Guarantees: The system provides a meaningful error condition.
Trigger: The mechanic's request at the counter.

Figure 19: Sample Use Case – Precision Level 1 – Scope

Use Case Analysis

You can see that several pieces of information have been added to the use case.

- Scope – The system is called the Automated Tool Tracking System or ATTS for short.
- Stakeholders – The roles that have a vested interest in this use case are listed with a brief definition of their interest.
- Precondition – The state or situation at the start of this use case. Think of it as the things this use case relies upon at its beginning.
- Successful Post Condition – The state that will exist when the use case is completed successfully.
- Minimal Guarantees – The minimal results to which the system can commit, the least it can do.
- Trigger – The event that acts as the catalyst to start the use case.

You might be inclined to move right into the Main Success Scenario (Precision Level 2) from here. However, move on to the next use case instead. Keep moving on through the list of use cases, filling in Precision Levels 0 and 1 only, until all use cases have been documented to this level.

It is important for the team to keep its thoughts on the total forest at this stage and not get stuck on any one tree. Keep the team focused on the broader set of use cases so they can make decisions about how use cases relate to each other. It is quite normal for you and your team to delete some use cases at this point, to merge some with others, or to add some that were overlooked as you work through the

> **It is important for the team to keep it's thoughts on the total forest at this stage and not get stuck on any one tree.**

whole set. It is more efficient to discover these changes early on before you invest too much time in any one use case.

This first pass through the use cases goes pretty quickly. You should not be surprised to see the team hammer through three or four use cases in an hour. They might get through even more if the team is experienced, they have a high degree of consensus, and the use cases are straightforward. You might even want to consider time-boxed discussions at 15 or 30 minutes per use case initially. Think about this when you try to figure out how many workshop meetings you will need.

Keep the use case register updated with the deletions, mergers, and additions as you go, too. You will finish your first pass when all use cases have Precision Levels 0 and 1 defined and when the use case register has been updated.

The Second Pass

In the second pass, you and your team define the Main Success Scenario for each of the use cases. Things go slower in this pass. Your workshop discussions get more involved. Design considerations enter the picture. You might create some low-fidelity user-interface (UI) prototypes to illustrate the steps in the Happy Path. You find that requirements, business rules, and business issues pop up in the discussions. You need to take the time to record these as you go so the thoughts don't get lost.

You need to learn to flip back and forth quickly from one display to another in your workshops. You haven't read about some of these just yet (stay tuned), but you will find yourself wanting to jump from one use case to another, to the use case register, to the UI mockups, to the requirements register, the business rules register and the issue log— all in the course of a second-pass workshop.

This enables you to capture all of the thoughts as they arise in discussions, which can become lively at times. Expect to pace the discussions so you can get the information recorded correctly.

In the second pass, as before, you should work through the collection of use cases in whatever order makes sense to you and your team. As each use case arises, you should document the Main Success Scenario of the use case. The Main Success Scenario should represent a typical sequence of actions that lead the actor to achieve the goal. It should be straightforward, easy to understand, and free from errors or branches in its logical progression.

When you have defined the Happy Path for your current use case, you should also take a few minutes to list the alternate paths and exceptions that you believe are required for the use case. However, try not to let your team define all of their steps at this point. As with the first pass, the team should complete this "Happy Path Plus" level of detail for all use cases before driving out any further detail in any of the use cases.

Figure 20 provides the second pass level of detail for our running sample use case to check out a tool to the flight-line mechanic.

Developing Use Cases

PRECISION LEVEL 2 – Main Success Scenario (MSS)
1. The mechanic requests a tool.
2. The operator retrieves the tool from its assigned storage location and registers as the operator of the transaction.
3. ATTS registers the operator and requests the mechanic's identification.
4. The operator records the mechanic's identification number.
5. ATTS registers the mechanic as the recipient and requests the tool identification number.
6. The operator records the identification number of the tool.
7. The operator completes the transaction.
8. ATTS records the checkout transaction.

PRECISION LEVEL 3 – Alternate Paths
1a. The mechanic requests more than one tool.

PRECISION LEVEL 4 – Exceptions
2a. The tool is not available in the assigned storage location.
2b. ATTS does not recognize the operator as a valid operator.
4a. ATTS does not recognize the mechanic as an authorize tool recipient.
7a. ATTS shows the tool as being already checked out to someone else.
7b. ATTS shows the tool as requiring calibration in the next 2 days.

Figure 20: Sample Use Case – Precision Level 2 – Main Success Scenario

You will notice that three new sections have been added to this use case.

Precision Level 2 - Main Success Scenario (MSS)

The basic course through a check-out transaction is documented in eight steps. The mechanic requests a tool. The operator retrieves it and then records the transaction in the ATTS system. The use case completes successfully and the mechanic is off to work.

Some guidelines about the Happy Path that you and your team should keep in mind would include (Cockburn 2001):

- The Happy Path should contain three-to-nine steps. If this is not the case, it might be best to break the use case into two-or-more smaller use cases. However, if the use case needs to be more or less steps than this range, so be it.
- Write your steps in complete sentences.
- Write your sentences with simple grammar and vocabulary.

- Write in active voice. Show "who has the ball" with the actor as the subject in the sentence.
- Show the process moving forward. This means no branches. The Happy Path should be the simplest and most direct way to achieve the goal. Employ alternate paths for branches.
- Show the user's intent, not the movements. This helps you to avoid the common temptation to build user interface design into the use case. For example, say, "The operator records the mechanic's identification number" rather than, "The operator scans the mechanic's flight line badge."
- Employ positive, goal-oriented action verbs. Instead of, "Check whether," try "Validate."

Precision Level 3 – Alternate Paths

Figure 20 shows one alternate path in the use case, which is that the mechanic requests more than one tool. Responding to this request will require some different processing logic, but don't record that logic just yet. At this point, just list the path.

Notice the number of the alternate path is "1a." The number, "1" tells you that the alternate path emanates from the first step in the Happy Path and the "a" tells you that this is the first alternate path that emanates from the first step in the Happy Path.

Precision Level 4 – Exceptions

In the example shown, the team has identified five potential exception conditions related to this use case. Once again, notice that the numbering of the exceptions tells you where the exception occurs and the number of the exception that emanates from the corresponding step in the Happy Path. For instance, "7b" marks that this is the second exception emanating from step 7.

Once again, the processing required to respond to the exception is not defined at this point. That detail will be added later. When it is added depends on the project lifecycle model as discussed under "Planning for Use Case Analysis" later in the book.

More on the Second Pass

Expect the team to have a lot more discussion on each use case in this second pass of analysis. You team members have different perspectives on the discussion based on their function. A business analyst thinks of additional functional requirements and business rules. A quality assurance analyst thinks of implications for test cases. A usability analyst will think in terms of user interface implications. Technical architects or developers think of class models or system integrations. Finally, the project manager might identify business issues.

You may find it very helpful at this stage to engage in some design discussions so that the team can work holistically. Some people work better within their "right brain" and want to see the user interface as the use case is defined. The addition of a user experience (UX) analyst to the team is good because the conversation is better thought out when good user interface design considerations are included in the mix.

> *Some people work better within their "right brain" and want to see the user interface as the use case is defined.*

You should allow time to have these added discussions and to capture the conclusions along the way. As before, expect to further consolidate and groom the list of use cases in the use case register as you go. Because of all these additional wrinkles, it might be best to have your solution definition workshops in the morning and then release your team members to work on issues, UI mockups, and requirements in the afternoon.

These aligned activities are discussed in the following subsections.

- User Interface Prototypes
- Functional Requirements Definition
- Business Rule Definition
- Non-Functional Requirements Definition
- Issue Research and Resolution

User Interface (UI) Prototypes

Some people think in words and some think in pictures. You should involve both types of people to produce better use cases and more appealing designs. Your customers will get a better feel for the system and your whole team will get a more holistic view. UI prototypes created while initial use cases are defined help users formulate and refine their expectations of the system under discussion. You can also reduce the risk of rework that results if you wait for all use cases to be defined before you start design work (Wiegers and Beatty 2013).

You should consider imbedding the UI prototypes right into the use case, either as pictures, objects or hyperlinks in the Supplemental Information section (Precision Level 5).

You should consider UI prototypes to be preliminary at this point because you will not be defining all of the alternate paths and exceptions yet.

UI Prototypes come in several flavors: whiteboard drawings, storyboards, wireframes, task flow diagrams, page flow diagrams, dialog maps, and mock-ups of the web pages. You should consider UI prototypes to be preliminary at this point because you will not be defining all of the alternate paths and exceptions yet. Therefore, stick to the less formal types of prototypes, such as storyboards or notional wireframes.

As mentioned before, you should hook an LCD projector up to a computer in a conference room to project the prototypes on the wall or overhead projector screen. This allows all participants to view them at once, increases participation, and accelerates the buy-in of the results. You can flip back and forth between use cases and UI prototypes to ensure their consistency with each other.

You might also consider building screen-flow models, sometimes called dialog maps or page-flow diagrams, before the individual page prototypes. A page-flow diagram is illustrated in Figure 21. This can be very helpful for keeping the individual pages in the context of the overall user experience.

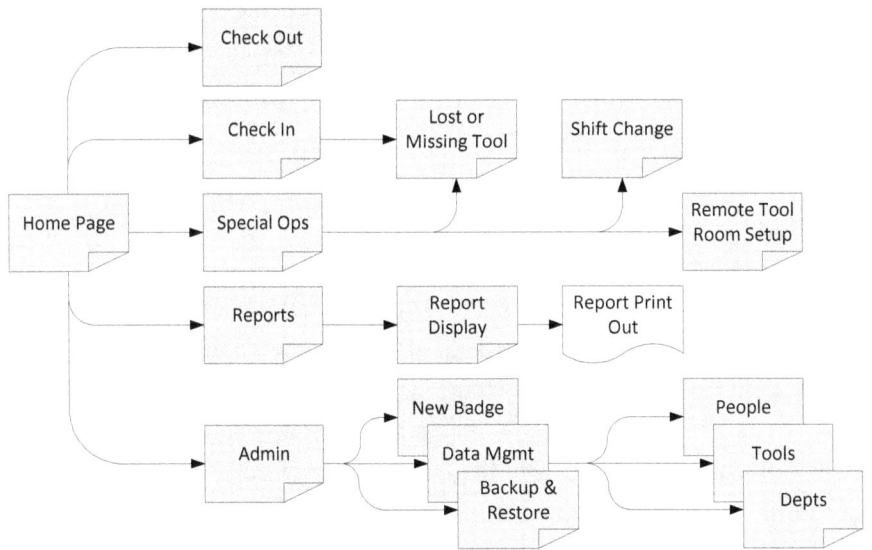

Figure 21: Page-Flow Example

Use Case Analysis

You can then develop page mock-ups of the individual pages as in Figure 22 for all pages or those pages where questions arise.

```
ATTS – Automated Tool Tracking System
   Checkout

         Enter or scan operator's badge number    [    ]

         Enter or scan borrower's badge number    [    ]

         Enter or scan tool barcode number        [    ]

         Enter or scan DONE                       [    ]

                                        ( Home )  ( Help )
```

Figure 22: Page Mockup Example

Functional Requirements Definition

A question that often comes up when a team works through use cases for the first time is, "Do use cases take the place of requirements?" No, they don't. They overlap, but they are different.

Functional requirements are statements that describe capabilities of the system required by the user while use cases describe how the system operates and interacts with the user.

Consider the running example use case to check out a tool. Let's say your team is defining the steps involved when someone pipes up, "Oh yeah, the system needs to check to see if the tool is due for periodic maintenance before anyone can check it out." This is a

Developing Use Cases

good example of how use case analysis helps the team think of requirements. It becomes an exception in the use case but it also spawns several requirements for the system. For example:

- The system must identify which tools require periodic maintenance.
- The system must be able to define separate periodic maintenance cycles for different individual tools.
- The system must prevent check-out of tools that are scheduled for periodic maintenance within the next 48 hours.

In other cases, you might think of a requirement that affects several different use cases, such as, "The system needs to be localized for German, Spanish, and French languages."

As the use cases are defined, the team captures requirements that arise in some sort of Requirements Register, as illustrated in Figure 23. That way, they won't be forgotten when the team moves on.

Rqmt ID	Description	Category	Priority	Use Case ID(s)	Test Case ID(s)	Remarks
R001	The system must identify which tools require periodic maintenance.	Check Out	High	UC01		
R002	The system must be able to define separate periodic maintenance cycles for different individual tools.	Check Out	High	UC01		
R003	The system must prevent check-out of tools that are scheduled for periodic maintenance within the next 48 hours.	Check Out	High	UC01		
R004	The system must force a Lost/Missing Tool Report if all tools are not accounted for at Check-In.	Check-In	High	UC02		
R005	The system must require a situation review at each shift change.	Special Operations	High	UC05		

Figure 23: Requirements Register

Assign a unique identity number to each requirement in the register so it can be referenced elsewhere. Other attributes can also be helpful such as priority, criticality, functional category, related test case identity numbers, and related use case identity numbers.

Non-Functional Requirements

Non-Functional requirements are also valid requirements but they are not user-oriented. Suzanne Robertson and James Robertson (Robertson and Robertson 1999) categorize non-functional requirements as follows:

- Look and Feel Requirements
- Usability Requirements
- Performance Requirements
- Operational Requirements
- Maintainability and Portability Requirements
- Security Requirements
- Cultural and Political Requirements
- Legal Requirements

You will find that non-functional requirements arise in use case discussions as well as functional requirements. You can list them in the same requirements register or in separate registers as you and your team prefer.

Most non-functional requirements are defined by a separate activity other than use case definition. For instance, your architect or development lead may author some sort of technical specifications document to focus on the technical design of the system. The nature and extent of the technical analysis varies from one project to the next but often involves some mix of these topics:

- Languages and tools.
- Configuration management tools and procedures.
- Development, test, and production environment specifications and promotion procedures.
- Systems and components involved.
- Databases involved and their design.
- Security requirements and specifications.
- Integration requirements and specifications.
- Infrastructure requirements and specifications.
- Deployment plans.

Non-functional requirements can be general and non-specific in the first stages of the project. You should refine them as the project moves into the construction stages.

Business Rule Definition

Business rules are specific guidelines, regulations, standards, specifications, or procedural requirements that must be adhered to in the operation of the system. You can think of them as miniature policy statements of your business owners about how the business operates. Types and sources of business rules include (Wiegers and Beatty 2013):

- Regulations
- Policies
- Data models
- Decisions
- Events
- Changes in the states of objects
- Restrictions or Constraints
- Computations

Business rules also arise while you and your team define use cases. Look for statements that sound like policies or rules and record them appropriately. You should keep a register for business rules, as you do for requirements and note them in the register when they come up. An example of a business rule register is shown in Figure 24.

BR ID	Short Name	Description	BR Category	Subject Matter Expert
BR001	All tools must be checked in at end of shift.	Before a mechanic can leave a shift, all tools checked out to that mechanic must be checked back into the Tool Room.	Policy	Betty Rubble
BR002	A missing tools report must be completed whenever a tool is lost.	If a tool is not turned in, it is assumed lost and its loss must be documented in a missing tools report before the mechanic can leave the work shift.	Policy	Betty Rubble
BR003	Only authorized individuals are allowed on the flight line.	The flight line is a restricted area, accessible only to those individuals that have a Flight Line badge assigned to them.	Security	Joe Kendricks

Figure 24: Business Rule Register

Several elements should be defined for each business rule.

- Unique Identifier (ID) – A unique number as defined by the team.
- Name – A short title for ease of reference.
- Description – Longer description to explain the rule and its applicability.
- Category – The type of rule or the source or basis for the rule.
- Subject Matter Expert (SME) – The person or department to contact about the rule.
- Related Use Case(s) – The ID(s) of the use case(s) affected by this business rule.

You and your team may choose to add, delete or change fields to describe the business rules as you see fit.

> **Business rules are specific guidelines, regulations, standards, specifications or procedural requirements that must be adhered to in the operation of the system.**

Some commonly asked questions about business rules are:

- How do they relate to business requirements? Business requirements often arise from business rules.
- Why do you keep them separate from the requirements? They are not the same thing. Requirements-focused capabilities are required by the system, whereas business rules reflect the policies of the business.
- How do they relate to use cases? They pop up in the discussion of use cases. You need to keep an ear open for the statement of a business rule, flip over to the business rule register to document it, and then flip back to the use case to continue the discussion.
- Why do you want to keep all of the business rules together in one register? They form the business policy basis for the system over its useful life and they grow, change, and go away over time. When you keep them together in one business rules register, you will see patterns and categories emerge for easier access, review, and management. Over time, without a business rules register, you and your organization will forget them and you or others will move on. Those left behind to maintain the system will wonder why you did what you did.

Issue Research and Resolution

You will also have "issues" arise in your conversations about use cases. Issues are questions or concerns about how something is done or differences of opinion about how things should be done. They occur throughout the project and you need to track them to ensure they get resolved, that the decisions are made, and that the resolutions are documented for future reference.

There will always be issues that can't be resolved on the spot. Someone will need to do some research or reach a decision with another group of stakeholders before it can be resolved. You need to record these issues in an issue log. Issues that are logged get worked. Issues that don't get logged remain unresolved. They might fester until late in the project, where they disrupt your schedules or result in unsatisfied customers.

An example of a project issue log is shown in Figure 25.

Issue No.	Issue Description	Date Posted	Issue Owner	Resolution	Date Resolved
I001	Is calibration the same thing as periodic maintenance?	4/21	Jerry	No, periodic maintenance means things like repacking wheel bearings.	4/24
I002	How do we check out a tool to a remote tool crib?	4/22	Frank		
I003	Do we need to fill out a Lost/Missing Tool Report for every single missing tool?	4/22	Jerry	Yes.	4/24
I004	Which browsers do we need to support?	4/23	Julie		

Figure 25: Project Issue Log

The steps involved in issue research and resolution are as follows:

1. Listen for issues that arise in the discussions as you engage in use case discussions. Examples include significant

questions that can't be answered by the team, differences of opinion that require escalation, or problems that need research.
2. When you recognize an issue, log it in the issue log and assign an owner. The owner is the person who will do the research or drive out a decision to resolve the issue. Make sure the owner agrees to research the issue and return with a recommendation in a responsive manner. Then, resume your discussions.

> *The owner is the person who will do the research or drive out a decision to resolve the issue.*

3. As discussions continue over the course of several days, keep the issue log in front of your team to ensure that issues get worked in a responsive manner.
4. When the issue owner presents a recommended resolution, review it with your team to ensure consensus and then log the resolution in the issue log to close out the issue.

Use Case Elaboration

Initial Use Case Definition provides you with a complete set of use cases for your project. They include the Happy Path and a list of the expected alternate paths and exceptions. However, there is still much work to be done. It will take you longer to define all of the alternate paths and exceptions than it took to complete the work done so far. That is simply because a lot of thought needs to go into them.

The timing of that additional work depends on the project life cycle you are following. This is discussed more under Planning for Use Cases, but worth a quick summary here. In a waterfall project, you should continue to drive out all the remaining sections of all use cases press in the requirements stage of your project. In an agile project, you should postpone elaboration of each use case until the sprint immediately preceding the sprint in which you intend to build out that use case (or begin building it out anyway). I like to call this "Just-In-Time Specification." It helps prevent specifications from becoming stale between definition and development. In addition, the developers who will build out the capability can contribute to the final specifications just before they build them.

> *It will take you longer to define all of the alternate paths and exceptions than it took to complete the work done so far.*

In the running example, Figure 26 presents the tool room checkout use case now fully dressed with alternate paths and exceptions.

PRECISION LEVEL 0 – Discovery
Use Case Name: Check out a tool to a mechanic
Primary Actor: Tool room operator
Narrative: A mechanic comes up to the tool room checkout counter to request a tool for the upcoming shift. The tool room operator retrieves the tool from its storage location and checks it out to the mechanic and records the complete transaction.
Alternate Primary Actors: Tool room supervisor
Context of Use: The operator is standing behind the counter of the flight-line tool room, which is usually caged behind protective screening. It is located in a restricted area, open only to people authorized to access the flight line.

PRECISION LEVEL 1 – Scope
Scope: Automated Tool Tracking System (ATTS)
Stakeholders and Interests:

Stakeholder	Interest
Mechanic	Wants the tool, in a hurry!
Tool Room Supervisor	Wants the information recorded accurately and quickly.

Precondition: The mechanic has approached the tool room counter and asked for a tool.
Successful Post Condition: The mechanic has the tool and the system has recorded the transaction accurately.
Minimal Guarantees: The system provides a meaningful error condition.
Trigger: The mechanic's request at the counter.

PRECISION LEVEL 2 – Main Success Scenario (MSS)
1. The mechanic requests a tool.
2. The operator retrieves the tool from its assigned storage location and registers as the operator of the transaction.
3. ATTS registers the operator and requests the mechanic's identification.
4. The operator records the mechanic's identification number.
5. ATTS registers the mechanic as the recipient and requests the tool identification number.
6. The operator records the identification number of the tool.
7. The operator completes the transaction.
8. ATTS records the checkout transaction.

PRECISION LEVEL 3 – Alternate Paths
1a. The mechanic requests more than one tool.
- 1a1. The operator retrieves all of the requested tools from their assigned storage location and registers as the operator of the transaction.
- 1a2. ATTS registers the operator and requests the mechanic's identification.
- 1a3. The operator records the mechanic's identification number.
- 1a4. ATTS registers the mechanic as the recipient and requests the tool identification number.
- 1a5. For each tool to be checked out:
  - 1a5(a). The operator records the identification number of the tool.
  - 1a5(b). ATTS checks the tool out to the mechanic and records the date and time.
  - End loop
- 1a6. Return to MSS step 8.

PRECISION LEVEL 4 – Exceptions
2a. The tool is not available in the assigned storage location.
- 2a1. The operator runs a query in ATTS to determine the most recent transaction for that tool.
- 2a2. The operator requests the mechanic to choose another tool.
- 2a3. Return to MSS step 1.

2b. ATTS does not recognize the operator as a valid operator.
- 2b1. The operator requests supervisor assistance.
- 2b2. The supervisor chooses to add a new operator.
- 2b3. Return to MSS step 2.

4a. ATTS does not recognize the mechanic as an authorized tool recipient.
- 4a1. The operator requests supervisor assistance.
- 4a2. The supervisor check the credentials of the mechanic and chooses to add the mechanic to the list of authorized mechanics.
- 4a3. Return to MSS step 4.

7a. ATTS shows the tool as being already checked out to someone else.
- 7a1. The operator checks in the tool.
- 7a2. Return to MSS step 6.

7b. ATTS shows the tool as requiring calibration.
- 7b1. The operator checks the tool out to the calibration center and puts the tool in the calibration bin to be forwarded to the calibration center.
- 7b2. The operator requests the mechanic to choose another tool.
- 7b3. Return to MSS step 1.

Figure 26: Fully Dressed Tool Room Use Case

Notice that the alternate paths and exceptions add a significant amount of content for the use case. It is normal for the effort required to dress out alternate paths and exceptions to exceed the amount of time necessary to define the earlier precision levels.

Also, notice the loop in the logic of alternate path 1a starting at step 1a5. Sometimes loops are necessary to repeat steps in a use case (Schneider and Winters 2001). In this case, the example shows a "for" loop, where a set of steps are repeated for a specified number of times. In this case, the number of times is determined by the number of tools to be checked out to the mechanic. In other loop scenarios, a set of tasks may take place while a certain condition holds true or until a condition proves true.

Sometimes, in agile projects where the amount of documentation is minimized, you might choose to finalize the use cases as user stories instead of added detail in the use case itself. In other words, you would not add the alternate path and exceptions definitions to the use case. Instead, you would list a set of user stories based on the information defined in the Initial Use Case definition step.

The Happy Path is addressed by a few user stories and each alternate path and each exception is a user story. For example, consider that functional capabilities implied in the tool room use case documented above might be represented by the stories listed in Figure 27.

> *Sometimes, in agile projects where the amount of documentation is minimized, you might choose to finalize the use cases as user stories instead of added detail in the use case itself.*

> 1. The operator is able to scan their barcode badge to register as the operator for the checkout transaction.
> 2. The operator is able to scan the requesting mechanic's barcode badge to register the mechanic as the recipient of the tools in this checkout transaction.
> 3. The operator is able to scan the barcode of the tool to check out the tool to the mechanic.
> 4. The operator is able to scan "Done" to complete the checkout transaction.
> 5. The ATTS is able to record the checkout transaction in the database.
> 6. The mechanic can check out multiple tools in one transaction.
> 7. The operator is able to use the mouse and keyboard as an alternative to the barcode scanner.
> 8. The operator is able to query the transaction history of a tool.
> 9. The supervisor is able to add a new operator to the system.
> 10. The supervisor is able to add a new mechanic to the system.
> 11. The operator is able to check in a tool.
> 12. The ATTS is able to detect if a tool is due for calibration and block its check out.
> 13. The operator is able to check out a tool to the calibration center.

Figure 27: Tool Room User Stories

This use case generated 13 stories. This is higher than normal but not all that unusual. A typical use case might generate nine user stories (three for the Happy Path plus three alternate paths and three exceptions), so 20 use cases might result in 180 user stories.

As mentioned earlier, another approach that you could take in an agile project is to dress out uses cases just prior to the sprint in which you intend to build out the use case. This provides a better historical document of the use cases for future reference over the useful life of the system.

The degree of formality you apply will vary from project to project and from company to company, depending upon their cultures. Do what you think fits the situation and what gets your team to a state of clarity about the users' functional requirements and how they will be addressed by the system(s) under discussion.

Pitfalls to Avoid

Use case analysis can be extremely helpful in software projects. Unfortunately, it can also be implemented poorly, just like any other practice. Here are some of the problems you should look out for and avoid as you put use case analysis to work.

Inadequate User Involvement

Use cases are intended to reflect the voice of the user. It is therefore surprising to see how often development teams engage in use case analysis without the participation of an actual member of the using community.

Too often, one or more of the team members think they know the users' needs well enough that they can act as a proxy for the using community. They drive the use case definition, making decisions for the users without consulting them. Cost or schedule savings are commonly expressed as a rationale. Another rationale is that there are too many users to consult, as in the creation of a mass-marketed software product.

> *"The greater problem is not the lack of knowledge, but the illusion of knowledge (paraphrased from Daniel Boorstin)."*

Sadly, this approach often fails. To paraphrase Daniel Boorstin, the American historian, "The greater problem is not the lack of knowledge, but the illusion of knowledge." The user proxies often create a system that makes sense to them, but confuses the users and fails to meet their needs. The result is disgruntled users, costly redesign, and sometimes a failed product.

To avoid this problem, arrange participation from the using community as you define your use cases and design the system. If you can't get their direct participation, have them name a representative as their proxy and then review your results with them at key intervals to get their approval.

Listening to the Wrong Voice

A related problem is listening to and expressing the wrong voice. The wrong voice is any voice other than the user's voice. You see this when excessive attention is paid to the voice of the organization (as voiced in business process models) or the voice of management. It can also stem from defining the system from the inside out, as would be the case with the voice of the architect, the developer, or the testers.

Once again, include actual users in your analysis whenever you can to avoid this problem.

Inadequate Stakeholder Involvement

This is also related to the proxy problem. Stakeholders are people or organizations with a vested interest in the project outcomes. Representatives of stakeholder organizations should be included in key project decisions. The definition of the system through use case analysis is one such key decision. Be sure to get their review and approval of your work.

Inadequate Understanding

This pitfall relates to the team's understanding of use case analysis. Like most of us, you learn as you go in software projects. Your understanding of use cases may be based entirely on the way they were used in one of your previous projects. That might be good or that might be bad depending on how they were implemented in that project. You should take some time to research and read about use case analysis before deciding how to employ them on your project.

Unilateral Decision-Making

Sometimes, one role on the project speaks over the rest in defining the use cases. This is unfortunate because use case analysis is inherently cross-functional. The best work comes from the blending of perspectives among all of the disciplines involved in the software project. The weave is stronger than any one thread.

> **The weave is stronger than any one thread.**

Be sure to moderate discussions to avoid any one perspective monopolizing the discussions and the resulting decisions.

Treating Them like Stories

On the surface, use cases and user stories look similar. But inside, they are very different. The user story is intended to be small. It is a promise for a later conversation. The use case is the documentation of a larger conversation.

As discussed earlier in this book, there are times when that extra effort of putting the conversation in writing really pays off. An example would include large, complex projects with interacting systems. Consider the craftsman's proverb, "Measure twice, cut once." Whenever rework will be costly, it is worth it to analyze the work a little more deliberately.

Overkill

Use case analysis can be overdone. You can easily find yourself over-thinking use cases if you're not careful. Learn to recognize what is "good enough" to avoid analysis paralysis. Good enough means that the team has a shared level of understanding about the use case that enables them to:

- Design the user interface.
- Develop the code.
- Write the test cases to ensure that the system operates as intended.
- Document the system for the user.

> *Good enough will be different for each team and for each project.*

Good enough will be different for each team and for each project. Ask your team when a use case is good enough. Talk it through as a team and make a team decision.

Gold Plating

An alternate form of overkill is gold plating. Gold plating, in the context of software engineering, means working beyond the requirements.

You might think that an extra widget on a page would be icing on the cake. However, if it's not required, don't do it. The agile lexicon includes the acronym, YAGNI— You Aren't Going to Need It. Develop what is required. That will be difficult enough.

Throwing Them over the Wall

A use case that is written by one party and then thrown over the wall to another party, no matter how well written, will be less effective than a use case that has been developed in a cross-functional workshop. The discussions, negotiations, compromises, and differences of opinion that arise in cross-functional discussions add a tremendous amount of understanding for the team.

Use cases go a long way to helping the specifications provided to geographically dispersed teams. They are recommended for these situations. Nonetheless, you should try to include all perspectives in the room when the use case is written whenever you can. Including people in the room virtually will help, though nothing beats in-person interaction for richness of communication.

Diving Too Deep Too Soon

Recall the earlier discussion about a first pass that covers Precision Level 1 before diving into the deeper precision levels for any one use case. Ignoring that recommendation will add cost and schedule to use case analysis through extra rework.

You will find that the list of use cases is very dynamic in the early stages of analysis. Traversing the entire list of use cases at a high level allows you to uncover most of the duplicate use cases, unnecessary use cases, and use cases that need to be combined without investing the time and effort required to define the more detailed information first.

Pedantic Implementation

No single approach to use case analysis will serve the needs universally. No author will capture everything. No one person's experience will address all situations. Following any one prescription rigidly and pedantically is a mistake.

The important thing is to assist the team in developing a shared, clear understanding of how the system will operate.

> *No single approach to use case analysis will serve the needs universally.*

Not Defining Dependencies

Another potential source of rework in projects that employ use cases is developing them out of sequence. Take some time during your planning efforts to think about the sequential dependencies among the collection of use cases.

Not Defining Context

Related to dependencies is understanding the larger business process context of your use cases. Business process analysis takes place at a higher level than use cases. It is common for the analysis of business processes to take place prior to listing the use cases.

> *Sometimes, business processes exist only as undocumented tribal knowledge.*

However, sometimes those business processes exist only as undocumented tribal knowledge, passed among members of the organization only verbally as part of "ramping up" into a job.

Spend some time assessing the situation of your project in the context of business processes. You may need to do some spade work there before defining use cases.

Not Maintaining Them

Use cases remain relevant to system operations over the entire useful life of the system. You should add to them, delete them, merge them, and modify them over that entire period as the system evolves. Not doing so won't hurt you, but you will spend a lot of time recreating the wheel if you don't. Also, as people move on, the reasoning for decisions made in the development of the system move on with them. Use cases capture much of that reasoning and can minimize the negative impact of turnover.

PLANNING FOR USE CASE ANALYSIS

Now that you have had a light introduction to the definition of use cases, let's look at some considerations for development of the project plan and the prioritization of use cases. To provide context for these thoughts, consider that most every project shares a generic life cycle such as the one shown in Figure 28.

Figure 28: A Generic Project Life Cycle Model

These three simple stages show the birth, life, and retirement of a project.

- **Project Commissioning** – Commissioning refers to the genesis of the project. Every project has some sort of process through which the project is defined, authorized, funded, and staffed.
- **Project Delivery** – The approved project is delivered in some fashion, whether it is with agile, waterfall, or somewhere in between.
- **Project Closure** – When the project has been completed, there are some things that should be done to close things

down in a well-ordered fashion and turn over the results to those who will support them.

Some aspects of use cases take shape in all three of these stages.

Use Cases in Project Commissioning

Commissioning may revolve around internal project selection processes or involve a sales cycle in the case of consulting projects.

The authorization of a project should get more attention than it does. This is where the first expectations are defined by the stakeholder community. Your project must meet those expectations to be considered successful.

It's also the first time that a project's cost and schedule are estimated. Recall the earlier discussions about the Cone of Uncertainty. Use cases can reduce the uncertainty and increase clarity of the system at the very start of project visualization. A one or two hour use case discovery discussion about actors and goals can provide several benefits.

- It can make a business case more attractive to the portfolio planning committee by providing a clear picture of intended system capabilities.
- It can provide a better basis of estimate for the development costs and schedules.
- A list of candidate use cases provides a good basis for prioritizing the capabilities to be built into the system.
- Development estimates based on candidate use cases reduce the business risk for both producer and customer alike.

Estimating with Use Cases

With your list of candidate use cases available, you can readily develop a high-level cost estimate. You can do this in a simple spreadsheet that lists the use cases and includes a column for each role that will be involved in the definition and build-out of the use cases. You can even add the columns for the roles right into the use case register. An example is shown in Figure 29.

UCID	Goal (Use Case Name)	Business Analyst	Solutions Architect	Usability Analyst	Developer	Quality Assuranc	TOTAL
UC01	Check out a tool	16	10	24	120	48	218
UC02	Check in a tool	12	2	12	96	38	160
UC03	Assign a new badge	8	6	16	80	40	150
UC04	Submit Missing Tool report	24	4	16	100	40	184
UC05	Conduct shift change	8	4	16	72	30	130
UC06	Refresh open tools list	0	8	0	32	16	56
	TOTALS	68	34	84	500	212	898

Figure 29: High-Level Estimating with Use Cases

Each member of your team estimates the number of hours necessary for each use case. This estimate will be much better than an estimate created without the list of use cases because use cases provide a much more concrete set of objectives.

A twist on this basic approach, a little more tailored to agile, is shown in Figure 30.

Use Case Analysis

Hours Per Typical Use Case			Business Analyst	Solutions Architect	Usability Analyst	Developer	Quality Assuranc	TOTAL
Hours per Typical Story			1.0	0.2	1.0	16.0	6.0	24.2
Avg. Num. of Stories per Use Case					9.0			
Hours per Typical Use Case			9.0	1.8	9.0	144.0	54.0	217.8
UCID	Use Case Name	Complexity						
UC01	Check out a tool	1.0	9.0	1.8	9.0	144.0	54.0	217.8
UC02	Check in a tool	0.8	7.2	1.4	7.2	115.2	43.2	174.2
UC03	Assign a new badge	0.7	6.3	1.3	6.3	100.8	37.8	152.5
UC04	Submit Missing Tool report	0.5	4.5	0.9	4.5	72.0	27.0	108.9
UC05	Conduct shift change	0.8	7.2	1.4	7.2	115.2	43.2	174.2
UC06	Refresh open tools list	0.4	3.6	0.7	3.6	57.6	21.6	87.1
		TOTALS	37.8	7.6	37.8	604.8	226.8	914.8

Figure 30: High-Level Estimating with Use Case Complexity

In this case, you estimate the number of hours for a typical user story and then multiply those estimates by the average number of stories per use case to derive an estimate for a typical use case. Then, you adjust the typical estimate for each individual use case with a complexity factor that defines the complexity of each use case relative to the average.

Of course, you should follow the method you think will provide the more accurate estimate given the information and people available.

This information can also be useful for determining the number of people required for each skill set and the number of sprints needed for an agile project. With this information, the schedule estimate is based on projected capacity for each skill set.

Managing Scope Change with Use Cases

Your list of candidate use cases also provides an excellent baseline against which to identify scope changes as the project progresses. This works well if you include an assumption in your statement of work similar to the following passage.

> "We have derived a presumptive list of use cases for the project based on our current understanding of the project requirements. Our estimates are based on this list. If any use cases are added, changed, or deleted from this list, we will need to follow prescribed scope change management procedures to evaluate possible adjustments to the cost and schedule estimates for the project."

This helps you preserve the integrity of the performance measurement baseline (scope, cost, and schedule) of the project. Without this clear definition of functional scope, the progressive elaboration that takes place in a project can lead to scope creep and overruns in budget and schedule— as well as profitability in contractual scenarios.

Use Cases in Waterfall Project Delivery

The life cycle model you follow on your project determines when you will define the advanced precision levels for your use cases. If you follow a waterfall life cycle model, you will define all six precision levels for all known use cases in the Requirements Definition stage. The software development life cycle of a typical Waterfall project is shown in Figure 31.

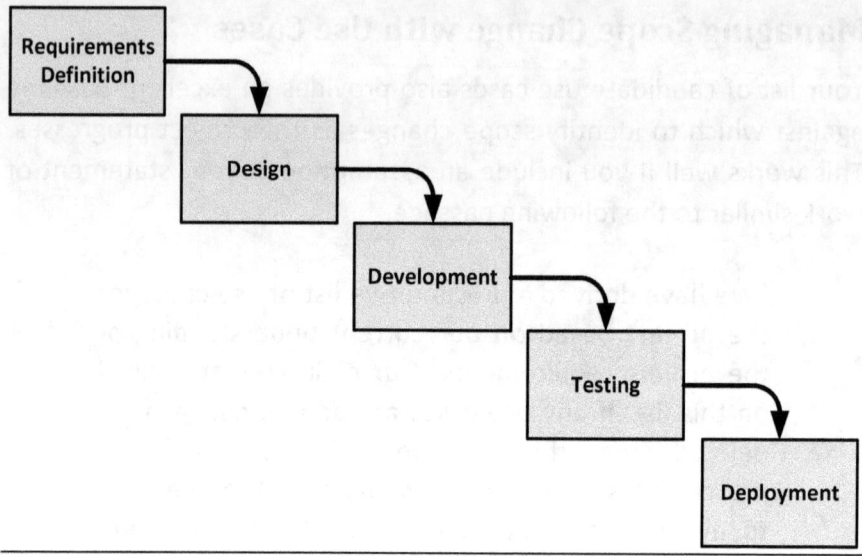

Figure 31: Waterfall Life Cycle

Sometimes, you might postpone part of this analysis to the Design stage. More commonly, however, you complete your use case definition in Requirements Definition and design the user interfaces in the Design stage.

Development takes place after the designs have been completed, testing takes place after development has been completed, and deployment takes place after testing is completed. Rework can become expensive in waterfall projects if requirements or design changes are identified in downstream phases.

Figure 32 shows the artifacts that are produced in the project mapped to the roles responsible for them and to the waterfall stages of the project. You can see why it's called waterfall, as artifacts cascade into others stage by stage.

Planning for Use Case Analysis

	Requirements	Design	Development	Testing
SA Solution Architect	Technical Specifications			
BA Business Analyst	User Analysis, Requirements, Candidate Use Cases, Use Cases			
UX User Experience Analyst	Page/Task Flow Diagrams	UI Prototypes, Site Map, Design Comps, Graphic Elements		
TW Technical Writer		Content Strategy, Content	Documentation	
DV Developer			Program Code, Build Scripts, Unit Test Scripts, Deployment Scripts	
QA Quality Assurance Analyst	Test Strategy	Test Plan		Functional Test Scripts, Defects

Figure 32: Waterfall Artifact Map

A question that is often asked by teams working in a waterfall model is, "Should we go back and update the use cases if they change in the design or development stages?"

My answer is, "Yes, you should go back and update the use cases whenever the system definition or operation changes." Use cases live on longer than the project in which they are developed. Other teams will use them as the system evolves through a series of new releases. The library of use cases grows and changes along with the system. Team members of new release projects start with use cases from previous projects and modify them as needed to accommodate new features or capabilities. The consumers of your use cases might modify them several years after you first defined them. They will benefit from your work, so make sure you give them use cases that are up to date, accurate, and complete. Note that this concept applies regardless of the project life cycle model. It is just as valid in agile projects as it is in waterfall projects.

> *New release projects check out use cases from the library and modify them to reflect the new features or capabilities.*

Use Cases in Agile Project Delivery

If you are working in an agile life cycle model, consider defining Precision Levels 1 and 2 for all use cases in the initial definition sprint that defines the project backlog. Complete Precision Level 1 in your project initiation or project charter task. Then, complete Precision Level 2 in a series of solution definition workshops. You should also list those alternate paths and exceptions that come to mind as part of this initial activity. Doing so helps you to better anticipate the complexity up front and to better define the user

Planning for Use Case Analysis

stories that are decomposed from the use cases. Completing this level of analysis goes quickly and leads to a much better backlog.

As you bring this initial definition sprint to a close, define the list of stories that make up the project backlog. This will include both technical stories, derived from technical specifications, and user stories derived from the use cases.

After you have defined the project backlog, you should look beyond the upcoming sprint to plan the content for all remaining sprints in the project. This activity estimates the number of sprints required to complete the project. Prioritize your use cases and their associated stories based on business value and then target them to specific sprints. The number of use cases you can tackle in a given sprint depends on the complexity of the use cases and your development capacity.

You may want to complete Precision Levels 3 – 5 just prior to developing those use cases. This practice of just-in-time specification is referred to as "Pipelining" as shown in Figure 33. The tasks for elaboration and build-out are defined in the sprint planning session for each of those sprints.

Sprint 0	Sprint 1	Sprint 2	Sprint 3	Sprint 4
UC1 Elaboration	UC1 Build-out	UC1 Test/Fix		
UC2 Elaboration	UC2 Build-out	UC2 Test/Fix		
UC3 MSS	UC3 Elaboration	UC3 Build-out	UC3 Test/Fix	
UC4 MSS	UC4 Elaboration	UC4 Build-out	UC4 Test/Fix	
UC5 MSS		UC5 Elaboration	UC5 Build-out	UC5 Test/Fix
UC6 MSS		UC6 Elaboration	UC6 Build-out	UC6 Test/Fix
			Test/Fix Overflow	Test/Fix Overflow

Figure 33: Pipelining Use Case Development and Testing

Use Case Analysis

In most flavors of agile development, you and your team make user stories your main focus rather than the use cases once you have defined your project backlog. Because they are smaller, user stories are more often prioritized and assigned to sprints rather than the use cases themselves. This makes sense, especially considering that a use case may be too large to complete within one sprint.

However, you may find it useful to be able to relate each user story back to its use case. Try employing a logical numbering scheme for your user stories that includes the use case ID. For instance, the first user story for use case UC03 becomes UC03A, the second becomes UC03B, and so on.

> **It can be useful to relate each user story back to its parent use case.**

Technical Stories, such as implementation of the test environment, won't conform to this convention because they don't relate to use cases. Instead, consider prefixing the serial numbers of technical stories with a "T" to designate technical stories, such as T014. Similarly, general stories, such as user documentation, might have a prefix of "G," such as G011.

In addition to naming stories with a logical naming convention, you might want to consider carrying it one step further to the numbering convention of the tasks defined to complete the stories. In sprint planning, when you define the tasks necessary to complete your stories, append a task number to the story number with a period (.) delimiter. For example, the second task associated with story UC03B would be UC03B.02. The third task for technical story T014 would be T014.03. This convention helps you and your team to keep track of the relationship of every task to its story and every user story to its use case. This can be very helpful in large projects that have thousands of tasks.

An agile approach implies some changes to the artifact map as shown in Figure 34. The efforts you undertake in the definition sprint create the level of detail required to build the project backlog. Elaboration of your use cases and the associated wireframes generates tasks to be accomplished incrementally, story-by-story, in the development sprints. Those elaboration tasks can include developers and occur just before the development of those stories. As a result, the specifications are fresh in the minds of the developer(s) when they write their code.

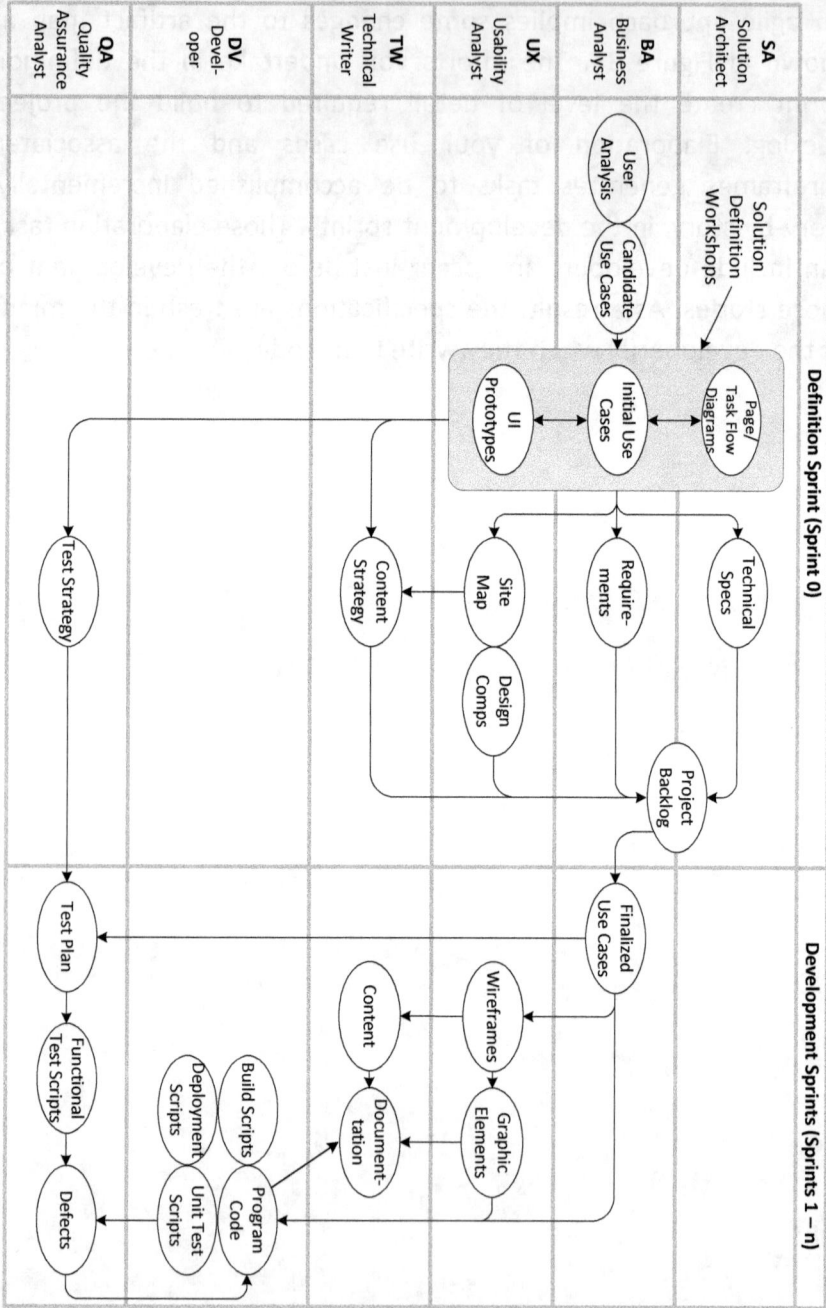

Figure 34: Agile Artifact Workflow

Notice in the artifact workflow that the Solution Definition Workshops include the initial use cases but that the final use cases are not completed until the development sprints. Even then, you would finalize the use case incrementally, one story at a time. The same holds true for wireframes. Create low-fidelity prototypes in the definition sprint, but create the final wireframes in the development sprints, one story at a time.

> *A Done List is a semi-standard list of tasks required to complete a user story.*

This incremental elaboration works well in your team's Done List. A Done List is a semi-standard list of tasks required to complete a user story. Defining what "done" means is a common requirement of agile methodologies. Creating a Done List as a miniature SDLC for user stories is one way to ensure that the meaning of "done" is consistently applied. An example is provided in Figure 35. The task number would be prefixed with the story number as discussed earlier.

.01 Finalize the use case portion for this story.
.02 Finalize the wireframe(s) for this story.
.03 Develop the code for this story.
.04 Develop unit test scripts for this story.
.05 Conduct unit tests for this story and remediate.
.06 Conduct code review for this story and remediate.
.07 Deploy code for this story to test environment.
.08 Develop the functional test script(s) for this story.
.09 Conduct functional tests for this story and remediate.

Figure 35: The "Done List"

This approach completes the fully dressed use cases and wireframes one story at a time, with each an increment to the end result.

Project Planning

This section describes a work breakdown structure for a sample project that employs use cases in a quasi-agile implementation. You can provide additional detail to this WBS to reflect the individual use cases for your project. You and your team can also tailor this WBS to reflect the needs of your specific project. This WBS is the same as the WBS in the responsibility matrix shown earlier in the discussion about roles and responsibilities. A Gantt chart view of the project plan is provided in Figure 36.

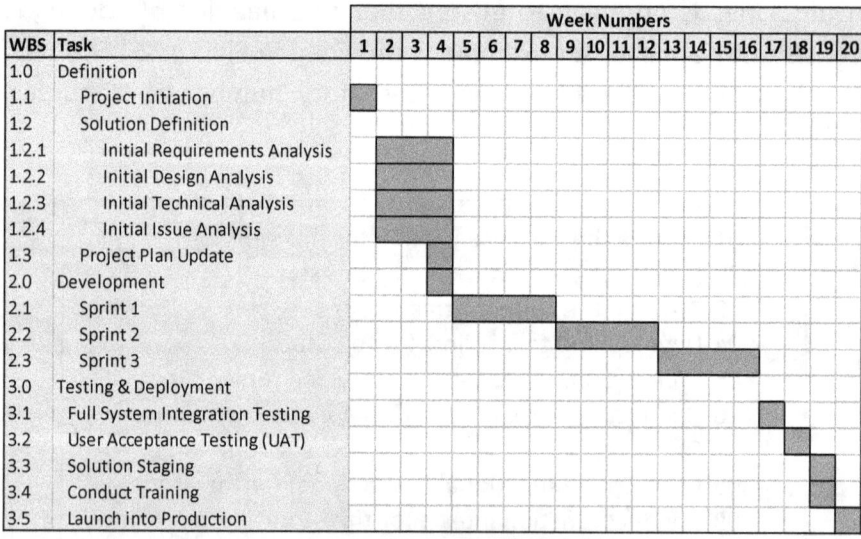

Figure 36: Gantt Chart

1.0 Definition

A definition sprint includes kicking off the project, defining the system to be provided, and preparing the project backlog.

1.1 Project Initiation

At the start of the project, you and your core team document a preliminary project plan in a Project Brief document. As part of this planning effort, you conduct your initial user analysis and use case discovery sessions. You also confirm resource commitments for the project as part of project initiation.

1.2 Solution Definition

Conduct a series of workshops with your core team, the product owner, and customer representatives to define the system that will be developed and delivered in this project. These workshops focus on four concurrent tracks of analysis.

1.2.1 INITIAL REQUIREMENTS ANALYSIS

Your requirements analysis starts with the candidate use cases that you listed in the use case discovery activity. In the workshops, you elaborate them into the initial use cases which include the Main Success Scenarios, lists of the anticipated alternate paths, and lists of anticipated exceptions. As you define these use cases, you also capture functional requirements and business rules as they arise, logging them into their respective registers.

1.2.2   INITIAL DESIGN ANALYSIS

Concurrent with the initial requirements analysis, the members of your team focus on the user's experience to develop initial design artifacts to coincide with the initial use case. An initial site map conveys the structure of the site to be developed (if there is one). The initial wireframes or user interface prototypes help the team to better understand the design of the system in discussion.

1.2.3 INITIAL TECHNICAL ANALYSIS

While the solution definition workshops are occurring, the Solution Architect and developers conduct related technical analysis. They write the Technical Specifications document and the initial non-functional requirements to help the team understand the technical dimensions of the system. This includes the system architecture, the technology platforms that will need to be put in place to support the system, and the technical capabilities that will need to be put in place for a successful system implementation.

1.2.4 INITIAL ISSUE ANALYSIS

Issues will arise as you work through the requirements, design, and technical analysis. Your product owner or other team members will need to research and resolve those issues as soon as possible so your team can build the system with confidence.

1.3 Project Plan Update

When you have completed the solution definition workshops, the technical analysis, and the issue resolution activities, you and your team can build the project backlog and make any updates needed to the project brief. The process of building the project backlog entails several tasks:

- Derive the user stories from the initial use cases and functional requirements.
- Derive the technical stories from the technical specifications and the non-functional requirements.
- List the general stories that are not covered by either user or technical stories, such as documentation.
- Estimate the effort necessary to complete each of the stories.
- Estimate the number of sprints required to complete the stories within the team capacity.

- Prioritize the stories and assign them to the sprints anticipated in the project.

Once you have defined the backlog, you are ready to begin development of the system in earnest.

2.0 Development

In the development sprints, you and your team build out the system, one story at a time. The number of sprints you need depends on your chosen sprint length, your development capacity (sometimes referred to as velocity), and the number of hours or story points you have estimated for the stories in the project backlog. Your team might choose to complete all of the stories for a use case before moving on to the next use case, or to follow a different strategy. For example, your team could choose to build out all the stories associated with the Main Success Scenarios of all use cases first and then pick up alternate paths and exceptions later on.

> *The number of sprints you need depends on your chosen sprint length, your development capacity (sometimes referred to as velocity), and the number of hours or story points you have estimated for the stories in the project backlog.*

In the Scrum model (Schwaber and Beedle 2002), you have a planning session, daily scrums, a demonstration, and a retrospection for each sprint. Within the sprint, each user story follows a somewhat repeatable set of tasks to build them out. As mentioned earlier, the Done List will include such tasks as:

- Fully dress the related use case for the portion covered by the story.
- Finish the design mock-ups or wireframes for the story.

- Develop the code for the story.
- Unit test and remediate the code for the story.
- Write the functional tests for the story.
- Conduct functional tests and remediation for the story.

Your development sprints continue until you have built out all the stories.

3.0 Testing and Deployment

Once you have built out all of the stories in the project backlog, you should include a testing and deployment sprint that includes several important tasks:

- Conduct Full System Integration Tests
- Conduct User Acceptance Tests (UAT)
- Stage the System
- Conduct Training (User and Support)
- Launch into Production

Prioritizing Use Cases and Stories

In the Project Plan Update task, you and your team should prioritize the user stories so that you work on the most important things first. In Mike Cohn's excellent book, "Agile Estimating and Planning" (Cohn 2006), he recommends that the prioritization of project work should be based on a combination of risk and value. Work of high risk and high value should be prioritized higher than stories of low risk and low value. Work of low risk but high value should be accomplished in the middle, and work of high risk and low value should be avoided.

While this is sound advice, I prefer to think of it a little differently. The value of a particular unit of work is a function of two things:

1. The benefit that is derived from the output of the work.
2. The risk that can be reduced by accomplishing the work.

These two components of value represent a trade-off decision that you must make with your customer. While your customer may initially perceive value to be based solely on the benefit derived, your team might more highly value the amount of risk that can be eliminated. For instance, your customer might think that including the shipping costs in a shopping cart application should be high on the priority chain, but your technical team might think that proving out the reference architecture should be high on the priority list.

> *While your customer may initially perceive value to be based solely on the benefit derived, your team might more highly value the amount of risk that can be eliminated.*

Conceptually, this trade-off can be represented in a diagram similar to Cohn's as illustrated in figure 37.

Use Case Analysis

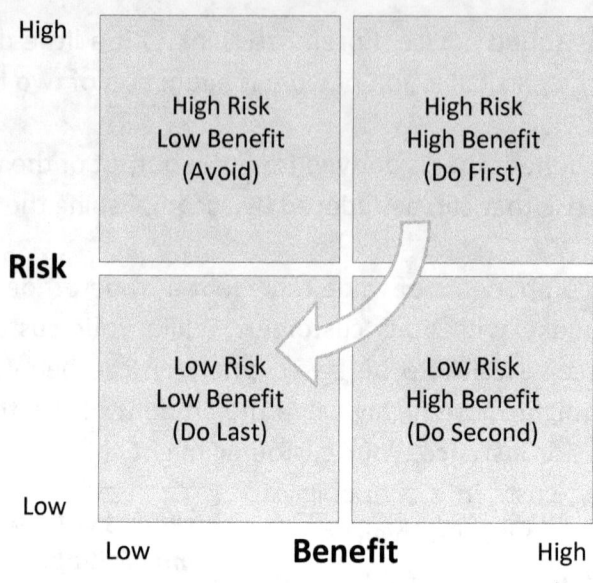

Figure 37: Value-Driven Prioritization

This implies that you should first build out those use cases that most reduce project risk and that provide the highest benefits realization.

In practical terms, you and your customer might select as highest priority those use cases which provide significant business value and involve the riskiest parts of the system architecture. Those first use cases will entail more work in the lower logic and data layers of the application than subsequent ones. You should plan that those first use cases will take longer and require more resources to complete than the use cases that are built out later in the project.

In large and complex projects, you might want to declare a separate Proof of Concept (POC) project to define and build out that initial set of use cases. This allows you to apply the knowledge gained from the POC to plan for the project which builds out the remainder of the use cases.

Use Cases in Project Closure

After you have developed all capabilities envisioned for the project, you move the system into a production environment and make the system available to the using community. As soon as this occurs, your project team is probably disbanded and scattered to the winds of future projects. However, you should try to get a bit more of their time to tidy up loose ends, archive the project artifacts, and arrange a well-ordered transition to the application support team(s) that will maintain the systems modified by your project.

Part of your clean-up process should include creating an archive of the use cases developed or modified in your project so that they can be employed by future project teams. Those future teams might want to modify, enhance, or otherwise build upon the work your team has done. Hopefully, your organization has an application manager, support team, or business analysis department that maintains use cases over time.

> *Part of your clean-up process should include creating an archive of the use cases developed or modified in your project so that they can be employed by future project teams.*

Too often, however, your project work on use cases is just copied to a secure storage backup area— never to be seen or heard from again. That's too bad, because your organization, if it lets that happen, misses an opportunity to improve its efficiency over time by maintaining libraries of use cases for the systems they support.

Use Cases in Application Support

You should maintain use cases over the useful life of the system(s) they describe. As future projects are initiated to enhance the systems, your use cases, built in the initial system construction, can save a great deal of time and anxiety to determine the incremental impact of the new project.

> *Your team members will come and go over time, so maintaining the use cases and other artifacts preserves their original intent and rationale.*

This same concept applies to the UI Mock-Ups, issue logs, business rules, and non-functional requirements.

Your team members will come and go over time, so maintaining the use cases and other artifacts preserves their original intent and rationale. They save time for downstream teams who would otherwise have to take a lot of time to reinvent these artifacts and figure out why things were done the way they were.

IN SUMMARY

I hope this somewhat light treatment of use cases adds value to your work and to your team. So much more could be said, but I have striven to provide enough to be useful without overkilling the subject with every possible nuance.

You might find it useful to share this material with your customer and your team members before commencing a project in which you are considering the implementation of use case analysis. As you have now read, there are many options to consider. Having this material as a common point of reference will help guide the decisions that you need to make.

Use case analysis is a powerful tool that every interactive software project should give serious consideration to. In today's agile development world, they are often forgotten in favor of the trendier user story concept or they are ruled out as too documentation-intensive. Unfortunately, I have seen too many instances where failure to employ them results in late discovery of requirements, costly rework, and scope creep.

If you want to read more about use cases, there are several excellent resources in the References section. Don't be surprised if what you read is not wholly consistent with the content of this book. That is the nature of the craft. You will find different approaches revolving around central themes rather than black and white consistency throughout the industry.

Please let me know what you thought of this book or ask any questions you might have by emailing me at davebly@pm-insights.com. I would love to hear from you about how this book has helped or how it could be improved. Good luck to you in your endeavors.

REFERENCES

Agile Manifesto. 2001. www.agilemanifesto.org

Armour, Frank and Miller, Granville. 2001. *Advanced Use Case Modelling: Software Systems.* Boston: Addison-Wesley

Bittner, Kurt and Spence, Ian. 2003. *Use Case Modeling.* Boston: Addison-Wesley

Boehm, Barry W. 1981. *Software Engineering Economics.* Upper Saddle River: Prentice Hall PTR

Cockburn, Alistair. 2001. *Writing Effective Use Cases.* Boston: Addison-Wesley

Cohn, Mike. 2004. *User Stories Applied for Agile Software Development.* Boston: Addison-Wesley

Cohn, Mike. 2006. *Agile Estimating and Planning.* Upper Saddle River: Prentice Hall Professional Technical Reference

Jacobson, Ivar, Booch, Grady, and Rumbaugh, James. 1999. *The Unified Software Development Process.* Reading: Addison-Wesley

Jacobson, Ivar with Christerson, Magnus, Jonsson, Patrick, and Övergaard, Gunnar. 1992. *Object-Oriented Software Engineering: A Use Case Driven Approach.* Harlow: Addison-Wesley

Larman, Craig. 2004. *Agile & Iterative Development: A Manager's Guide.* Boston: Addison-Wesley

Leffingwell, Dean. 2011. *Agile Software Requirements: Lean Requirements Practices for Teams, Programs, and the Enterprise.* Upper Saddle River: Addison-Wesley

Leffingwell, Dean and Widrig, Don. 2003. *Managing Software Requirements: A Use Case Approach, Second Edition.* Boston: Addison-Wesley

McConnell, Steve. 1998. *Software Project Survival Guide: How to Be Sure Your First Important Project Isn't Your Last.* Redmond: Microsoft Press

McConnell, Steve. 2006. *Software Estimation: Demystifying the Black Art.* Redmond: Microsoft Press

Övergaard, Gunnar and Palmkvist, Karin. 2005. *Use Cases: Patterns and Blueprints.* Indianapolis: Addison-Wesley

Robertson, Suzanne and Robertson, James. 1999. *Mastering the Requirements Process.* London: Addison-Wesley

Schneider, Geri and Winters, Jason. 2001. *Applying Use Cases, Second Edition.* Boston: Addison-Wesley

Schwaber, Ken and Beedle, Mike. 2002. *Agile Software Development with Scrum.* Upper Saddle River: Prentice Hall

Wiegers, Karl and Beatty, Joy. 2013. *Software Requirements, Third Edition.* Redmond: Microsoft Press

APPENDIX A – USE CASE TEMPLATE

PRECISION LEVEL 0 – Discovery

Use Case Name: [The goal of the actor]

Narrative: [A brief, general description of the conversation.]

Primary Actor: [The person or system that has the goal.]

Alternate Primary Actors: [Other persons or systems that have the goal.]

Context of Use: [Actor's intent, motivation, experience and expertise; technical or environmental constraints.]

PRECISION LEVEL 1 – Scope

Scope: [The system under discussion.]

Stakeholders and Interests: [Those interested in the outcome, validators or rule providers.]

Stakeholder	Interest
(your text goes here …)	
(your text goes here …)	

Precondition: [The state of affairs that the system can rely upon existing prior to the use case.]

Success Post Condition: [The state of affairs that the system can rely upon after the use case is successfully processed.]

Minimal Guarantees: [The least that happens, such as an error message.]

Trigger: [The event or situation that kicks it off, or causes the actor to seek the goal. NOTE: there is not always a trigger.]

PRECISION LEVEL 2 – Happy Path

Main Success Scenario (MSS) – [List 3-9 steps; use complete sentences, simple grammar, show "who has the ball", show the process moving forward.]

1. (your text goes here ...)

    a. (your text goes here ...)
    b. (your text goes here ...)

2. (your text goes here ...)

3. (your text goes here ...)

PRECISION LEVEL 3 – Alternate Paths [Start with the MSS step at which the branch occurs. Use a path number for each branch. Define the response steps and tie back into a step in the MSS.]

Step#, Path#. (your text goes here ...)
Example: 2a. User clicks "More ..."

1. response text goes here ...

2. response text goes here ...

3. response text goes here ...

PRECISION LEVEL 4 – Exceptions [Start with the MSS step at which the exception occurs. Identify an exception number for each error. Define the response steps and tie back into a step in the MSS.]

Step#, Exception#. (your text goes here ...)
Example: 2a. User clicks "More ..."

1. response text goes here ...

2. response text goes here ...

3. response text goes here ...

PRECISION LEVEL 5 – Supplemental Information [Record any additional items of interest about this use case in this section.]

Parent Use Case(s) [Called from ...]

Child Use Case(s) [Includes ...]

Test Script or Testing Considerations

Screen shots, wireframes, UI prototypes, mock-ups, etc.

Technology and Data Variations List

Security Considerations

Validation:

1.

2.

3.

Notes From Discussion

1. ...

APPENDIX B – GLOSSARY

Actor
: A person, organization, or system that initiates a use case to achieve a goal of that actor. The Primary Actor initiates the use case. A Secondary Actor is a person or system involved in the conversation with the Primary Actor that responds to or carries the conversation forward to other systems or actors. A Tertiary Actor is a person or system triggered by a Secondary Actor to support the achievement of the Primary Actor's goal.

Agile Lifecycle Model
: A technique for software development that is characterized by incremental and iterative development cycles wherein a subset of the system is developed.

Alternate Path
: A branch in the conversation initiated by an actor's decision.

Build Scripts
: Programs written to assemble all of the programs into the full system.

Business Owner
: A member of the sponsoring business community that acts as a focal point for the project or team.

Candidate Use Cases	A preliminary set of use cases to be defined and built out in the project. They are considered candidates early in the project because of the high probability that they will be added to, modified, merged, split, or deleted in the discovery and elaboration process associated with use case analysis.
Context of Use	A description of the conditions under which the system is used.
CRM	Customer Relationship Management
Deployment Scripts	Programs written to package and move the system files to the desired location for a particular usage environment, such as QA, UAT or Production.
Design Comps	Comparative designs for a website to help the customers determine their graphic design or "look and feel" preferences.
Exception	An expected error condition that might be encountered in the course of the use case.
Extends	The relationship between two use cases where the second use case is very similar to an alternate path in the first use case.
FTP	File Transfer Protocol
Fully Dressed	Fully elaborated use case in the structured text format.
Functional Requirements	Needed features of a system that relate to the functional usage and capabilities of the system.

Appendix B – Glossary

Functional Scope	A definition of what the system will do and the services it will perform. It is often the customer's single most important measure of success for a software project.
Functional Test Scripts	Tests written to determine whether or not the system functions as required. Test cases tend to be conducted manually, whereas test scripts are sometimes automated to be run in a regression suite.
Graphic Elements	The various graphic files, such as .JPG or .GIF files that are applied to the web page to implement the desired graphic design.
Happy Path	Synonym for Main Success Scenario
Includes	The relationship between two use cases when the first use case calls the second use case.
Initial Use Cases	The initial partial drafts of the use cases that cover just the top precision levels.
Issue	A question, concern, or required decision that could affect the feasibility, cost, schedule, or scope of the project.
IVR	Interactive Voice Response unit. An element of the telephony for a call center that provides for menu-based choice selection and information collection over the telephone.

Main Success Scenario (MSS)

    The primary successful path of the conversation between the actor and the system that achieves the goal of the use case.

Minimal Guarantee   The least response provided by the system in the conversation, such as an error message of some sort.

MSS   Main Success Scenario

Non-Functional Requirements

    Statements of the capabilities of a system that relate to characteristics of the system other than its functional usage and capabilities, such as internal architecture, infrastructure, performance, security, compatibility, portability, integration mechanisms, and business continuity.

Persona   A profile of a person that uses the system or an actor that represents a class of user. The persona paints a picture of an individual who represents the typical skills, characteristics, and behavioral patterns of that class of user. It assists the designers of the system to create designs that will be intuitive to that class of user.

Appendix B – Glossary

Pipelining	A concept employed to sequence the activities in the life cycle of a story or use case which calls for the specification activities to take place in the iteration just prior to the iteration in which the capability will be developed. Sometimes, the tests and fixes for that capability further take place in the iteration after the iteration in which the capability is built.
Precondition	The state or condition of the system at the start of the use case.
Primary Actor	See Actor
QA	Quality Assurance
Scenario	Any one of several unique paths through the logic branches and exceptions in a use case.
Scope	A definition of the goals, objectives, deliverables or boundaries of a system project.
Secondary Actor	An actor that initiates conversations with other systems or actors to achieve the primary actor's goal.
Sequence Diagrams	A diagram that shows how processes operate with one another and in what order. Different processes or objects that are active at the same time are shown as parallel vertical lines and the messages exchanged between them are shown as horizontal arrows in the order in which they occur. This allows the specification of process scenarios in a graphical manner.

Site Map	A site map (or sitemap) is a list of pages of a web site organized in hierarchical fashion. This helps visitors and search engines find pages on the site.
Software Engineering	
	The field of endeavor focused on the full life cycle of definition, development, test, support, and management of software.
SOW	Statement of Work
Stakeholder	A person, group, or organization with an interest in a use case or project.
State	The present status of a system or entity; a snapshot of the measure of various conditions in the system.
Statement of Work	A somewhat formal definition of the work to be performed as part of the project.
Story	A small piece of system capability that has value to the user. Stories also cover non-functional requirements, such as infrastructure implementation. Stories can be independently estimated, developed, and tested. They are the level of work at which daily work is accomplished. Stories are not considered complete until they have been deployed, tested, and rectified.
Storyboard	Graphic illustrations or images displayed in sequence to visualize a sequence of events or activities, motion picture, animation, motion graphic, interactive media sequence, or website interaction.

Appendix B – Glossary

Successful Post Condition
: The state or condition that will be present after successful completion of the use case.

Sunny Day Scenario
: Synonym for Main Success Scenario

System Under Discussion (SUD)
: The primary system involved in the conversation in the use case.

Task Flow Diagrams
: Graphic illustrations of the tasks and sequences involved in work processes.

Tertiary Actor
: An actor that converses with a secondary actor to achieve a goal of the secondary actor.

Test Case
: A test case is a document which defines the steps necessary to test a portion of the software to ensure that it is fit for purpose. Test cases generally document manual testing steps.

Test Script
: A test script is very much like a test case except that the test script is generally an automated test that can be run without human intervention.

Traceability
: Maintenance of an audit trail that maps how requirements are addressed by system design specifications and subsequently translated into development efforts, test cases, and acceptance to ensure that all stipulated requirements are manifested in the system developed over the course of the system project.

Trigger	An event or condition that causes a use case to be started.
UAT	See User Acceptance Testing
UI Prototype	A preliminary layout of the page or pages involved in the use case.
Unit Test	A testing method in which the individual units of source code are tested to determine if they are fit for use. A unit is the smallest testable part of an application such as a function, procedure or method.
Use Case	A method of defining functional requirements by documenting the conversation between an actor and a system to achieve a meaningful goal of using the system. An actor can be a person, organization, or a computer system. These conversations have a give-and-take, act, and request-and-respond flow to them. It is this interaction that the use case captures.
User Acceptance	The process through which the user community reviews and accepts the system that has been developed to address their requirements.
User Acceptance Testing (UAT)	User acceptance testing is a stage in the overall testing of a system that focuses on securing the acceptance of the system from the using community for the system. Use cases provide an excellent foundation for this testing because the collection of use cases represent the agreed upon set of goals of the users. Test cases are often developed from

Appendix B – Glossary

                        the use cases, one test case for each scenario defined in the use case.

Waterfall Lifecycle Model

                        A project life cycle model characterized by a sequential software development process, in which progress flows downward (like a waterfall) through the phases of the project: Definition, Design, Development, Test, and Deployment.

WBS   See Work Breakdown Structure

Web application   A type of interactive system architected for the World Wide Web. This entails a web browser for the presentation of user-evident capability, a web server to present that capability to the web browser, and various other servers to manage the programs and data associated with the system.

Wiki   A website that allows the easy creation and maintenance of any number of interlinked web pages via a web browser with a simplified markup language or a text editor.

Wireframe   An illustration of the layout of fundamental elements in the user interface. Typically, wireframes are completed before final artwork is developed. Wireframes allow for the development of variations of a layout to maintain design consistency throughout the site. This is an important part of the initial development stage because it creates user expectations and helps develop familiarity with the site.

Work Breakdown Structure

A work breakdown structure is a deliverable-oriented decomposition of a project into smaller components. The work breakdown structure should be defined around the deliverables that will be produced by the project. Each of these deliverables should be decomposed into interim deliverables. Each interim deliverable should then have defined for it the tasks or steps that will be accomplished to produce the deliverable.

INDEX

acceptance, vii, 2, 133
actor, 1, 9, 10, 11, 19, 22, 23, 24, 25, 28, 37, 38, 49, 52, 55, 56, 57, 58, 59, 60, 61, 63, 64, 67, 68, 72, 74, 98, 123, 124, 127, 130, 131, 133, 134
agile, 2, 3, 5, 29, 32, 48, 52, 65, 86, 89, 90, 97, 99, 100, 104, 106, 107, 108, 109, 110, 114, 119, 121, 122, 127
Agile Manifesto, 3, 32, 121
alternate path, 11, 22, 24, 25, 28, 31, 32, 52, 66, 72, 74, 77, 85, 86, 89, 90, 104, 111, 113, 124, 127, 128
analyst, 15
application, 2, 32, 56, 57, 68, 115, 116, 117, 134, 135
architect, 8, 15, 44, 80, 112
architecture, 15, 36, 44, 112, 115, 116, 130
archive, 117
Armour, Frank, 15, 61, 121
artifact, 5, 16, 47, 63, 102, 107, 109, 111, 117, 118
authorization, 55, 98
BA, 44, 64

backlog, 65, 104, 105, 106, 107, 111, 112, 113, 114
basic conversation, 23
Basic Course, 11
benefit, 7, 35, 104, 115
Big Design Up Front (BDUF), 32, 65
Bittner, Kurt, 13, 15, 58, 121
Booch, Grady, 21, 121
brainstorming, 57, 60
branch, 52, 66, 72, 74, 131
browser, 37, 135
budget, 56, 101
business analyst, 7, 8, 28, 41, 42, 44, 48, 56, 66, 75, 117
business case, 98
business intelligence, 38
business logic, 30
business owner, 6, 14, 18, 43, 56, 81, 127
business process, 4, 27, 28, 29, 92, 96
business process modeling, 26, 27
business process models, 4, 28, 29

137

business rule, 5, 15, 28, 43, 44, 63, 64, 71, 75, 76, 81, 82, 83, 111, 118
business user, 21
business value, 5, 32, 105, 116
candidate, 37, 49, 51, 52, 55, 60, 61, 66, 98, 99, 101, 111, 128
candidate use cases, 56, 60, 128
capacity, 100, 105, 112, 113
category, 64, 80
charter, 21, 55, 104
class model, 2, 75
closure, 97, 117
coarse-grained stories, 29
Cockburn, Alistair, viii, 11, 22, 52, 73, 121
collaboration, 13, 64, 65, 66, 68
commercial, off-the-shelf software, 27
commissioning, 97, 98
component, 81, 115, 136
compromise, 94
computer system, 9, 10, 13, 16, 21, 30, 35, 134
conceptual integrity, 44
cone of uncertainty, 50, 51, 98
configuration management, 81
constraint, 81
content, 42, 67, 68, 89, 105, 119
content strategist, 42
context, 64, 96, 123, 128
contract, 32
copy writer, 42
cost, 1, 6, 33, 43, 44, 51, 62, 68, 98, 99, 101, 115, 129
COTS, 27, 37, 38
cross-functional, 6, 14, 93, 94
custom development, 37
customer, 1, 2, 3, 6, 7, 8, 15, 18, 32, 34, 50, 52, 55, 59, 65, 76, 84, 98, 111, 115, 116, 128, 129
customization, 38
data, 29, 30, 37, 38, 39, 59, 116, 135
data warehouse, 38
decompose, 32, 33
demonstration, 113
dependency, 96
deploy, 30, 33, 132
deployment, 81, 114, 128, 135
design, vii, 2, 7, 9, 12, 13, 14, 15, 16, 18, 24, 28, 29, 33, 36, 38, 44, 45, 50, 51, 55, 65, 66, 74, 75, 76, 80, 81, 102, 104, 111, 112, 113, 128, 129, 130, 133, 135
designers, 2, 13, 14, 15, 16, 130
desk procedures, 45
develop, 3, 5, 15, 30, 33, 55, 65, 94, 104, 111, 117, 127, 131, 132, 133, 134, 135
developer, 2, 7, 13, 14, 16, 32, 41, 43, 44, 45, 75, 86, 107, 112
development, vii, viii, 3, 7, 12, 14, 16, 18, 21, 29, 32, 33, 34, 37, 41, 43, 44, 45, 51, 52, 55, 56, 61, 66, 80, 86, 91, 96, 97, 98, 101, 102, 104, 105, 106, 107, 109, 113, 114, 119, 127, 132, 133, 135
diagram, 19, 20, 21, 22, 25, 28, 45, 76, 77, 115, 131

Index

differences of opinion, 35, 84, 85, 94
disciplined, 33
discovery, 1, 5, 34, 49, 51, 52, 55, 56, 57, 61, 62, 66, 68, 98, 111, 119, 128
document management, 65
documentation, 2, 3, 28, 32, 34, 89, 106, 112, 119
documenter, 2, 14
done list, 109, 113
DV, 44
e-commerce, 37
editor, 42
elaborate, 24, 51, 111
elaboration, 24, 51, 55, 85, 86, 105, 107, 109, 128
enhancement, 37
epic, 29
error, 11, 66, 72
error conditions, 11, 53
estimate, 1, 6, 18, 33, 47, 51, 62, 64, 65, 98, 99, 100, 101, 105, 113, 114, 121, 132
event modeling, 56, 59
events, 5, 26, 49, 58, 59, 60, 66, 68, 132
exception, 11, 19, 24, 25, 31, 32, 53, 66, 72, 74, 75, 77, 79, 85, 86, 89, 90, 104, 111, 113, 125, 128, 131
expectation, 1, 13, 33, 76, 98, 135
extend, 128
extract, transform, and load (ETL), 39

feasibility, 51, 129
first pass, 67, 71, 72, 95
flow chart, 19, 25
Flow of Events, 11
fully dressed, 3, 22, 23, 25, 31, 34, 54, 86, 88, 110, 128
functional fit, 6
functional requirements, vii, 2, 9, 10, 15, 35, 39, 75, 76, 78, 80, 90, 111, 112, 128, 134
functional test, 3, 43, 45
Gantt, 110
general story, 106, 112
geographically dispersed, 33, 95
goal, 1, 8, 9, 10, 11, 15, 16, 22, 23, 25, 30, 33, 38, 43, 56, 57, 59, 60, 61, 63, 66, 67, 68, 72, 74, 98, 123, 124, 127, 130, 131, 133, 134
gold plating, 94
good enough, 93, 94
happy path, 11, 22, 52, 66, 67, 71, 72, 73, 74, 85, 89, 90, 124, 129
hyperlink, 64, 68, 76
ID, 63, 64, 82, 106
include, 125, 129
information architect, 45
infrastructure, 81
initial use case, 129
instance, 22, 24, 29, 41, 57, 64, 74, 80, 106, 115
integration, 18, 30, 59, 75, 81
integrity, 101
interact, 35, 78

issue, 5, 14, 43, 44, 53, 63, 64, 71, 75, 76, 84, 85, 112, 118, 129
iteration, 32, 131
iterative, 29, 127
Jacobson, Ivar, viii, 21, 121
just-in-time specification, 86
language, 81
Leffingwell, Dean, 13, 21, 30, 35, 122
lexicon, 49, 94
life cycle, 51, 97, 101, 104, 131, 132
logic, 2, 11, 51, 52, 74, 89, 116, 131
loop, 89
low-fidelity prototype, 25, 109
Main Success Scenario, 11, 22, 23, 25, 28, 52, 66, 67, 70, 71, 72, 73, 124, 129, 130, 133
maintainability, 19, 80
maintenance, 16, 37, 62, 78, 79, 135
manager, 15
market research, 56
messaging, 37
method, 16, 51
methodology, 2, 34
middleware, 43
Miller, Granville, 15, 61, 121
minimal guarantee, 52, 70, 124, 130
mock-ups, 2, 53, 57, 76, 78, 113, 126
MSS, 11, 22, 32, 73, 124, 125, 130

narrative, 20, 21, 22, 25, 52
needs, 13, 34
negotiation, 94
non-functional, 39, 44, 48, 76, 80, 112, 118, 130, 132
normal course, 11
numbering, 74, 106
Övergaard, Gunnar, viii, 11, 121, 122
overruns, 2, 101
page-flow diagram, 77
Palmkvist, Karin, 11, 122
participation, 66, 77, 91
performance, 80
persistence-layer, 43
persona, 130
pipelining, 105, 131
pitfalls, 91
planning, 5, 27, 75, 85, 97, 110, 114, 121
planning session, 105, 113
PM, 43
PO, 43, 59
POC, 116
portability, 80
portfolio planning, 98
position, 41, 43, 58, 145
post condition, 70, 124, 133
precision, 24, 44, 52, 53, 55, 67, 89, 95, 101, 129
precision level, 52, 53, 63, 66, 67, 69, 70, 71, 73, 74, 76, 95, 101, 104, 105
precondition, 70, 124, 131
presentation layer, 30, 43
presumptive, 101

Index

prioritize, 33, 105, 113
priority, 64, 80, 115, 116
process, 3, 7, 21, 26, 27, 28, 29, 33, 34, 35, 36, 37, 38, 41, 45, 49, 51, 52, 53, 55, 59, 66, 67, 74, 96, 97, 112, 117, 124, 128, 131, 133, 134, 135
product manager, 7, 43
product owner, 6, 8, 43, 48, 66, 111, 112
profitability, 62, 101
progressive elaboration, 5, 20, 49, 52, 101
Project Brief, 21, 111
project life cycle, 3, 48, 85, 104, 135
project management, 49
project manager, 7, 8, 43, 75
project plan, 2, 65, 97, 110, 111
project type, 36
proof of concept, 116
prototype, 45, 76
proxy, 91, 92
pseudo code, 51
QA, 42, 45, 64, 66, 128, 131
quality assurance, 7, 14, 16, 29, 33, 41, 45, 75, 131
quality assurance analyst, 7, 14, 16, 41, 75
readability, 18
realization, 116
reference, 35, 82, 84, 90, 115
relational database, 65
request for proposal, 55
request-and-response, 23

requirements, vii, 1, 2, 3, 5, 7, 12, 13, 14, 15, 16, 33, 35, 38, 39, 43, 44, 45, 48, 49, 51, 55, 64, 65, 67, 71, 76, 78, 79, 80, 81, 82, 83, 86, 101, 102, 111, 112, 118, 119, 132, 133, 134
Requirements
   functional, vii, 2, 9, 10, 14, 15, 35, 39, 75, 76, 78, 80, 90, 111, 112, 128, 134
   non-functional, 80
requirements definition, 101, 102
requirements register, 79
responsibility, 41, 43, 44, 45, 47, 48, 58, 66, 110
responsibility matrix, 45, 46, 47
retrospection, 113
revision, 66
rework, 102
risk, 6, 76, 98, 114, 115, 116
role, 5, 7, 9, 27, 28, 41, 42, 43, 45, 47, 48, 57, 58, 70, 93, 99, 102, 110
role matrix, 42
Rosetta Stone, iii, iv, vii, 3, 14, 39
Rumbaugh, James, 21, 121
SA, 44
scalability, 18
scenario, 2, 3, 11, 12, 20, 31, 32, 52, 72, 89, 101, 111, 113, 131
schedule, 1, 6, 33, 43, 51, 62, 84, 98, 100, 101, 129
Schwaber, Ken, viii, 32, 113, 122

scope, vii, 1, 6, 18, 21, 33, 51, 52, 62, 67, 69, 70, 101, 119, 123, 129, 131
scope change, 18, 62
screen-flow model, 77
Scrum, viii, 113, 122
scrums, 113
SDLC, 2, 109
second pass, 67, 71, 72, 75
security, 39, 53, 80, 81, 126, 130
sequence diagram, 131
session management, 29
Software Development Life Cycle, 2
software engineering, viii, 9, 50, 121, 132
software product, 56, 91
solution, 1, 2, 7, 16, 39, 44, 52, 75, 104, 112
solution architect, 44, 45, 48
solution definition, 109, 111
solution strategy, 1
SOW, 56, 132
specialization, 42
specification, 7, 15, 48, 52, 55, 80, 81, 86, 105, 107, 112, 133
Spence, Ian, 13, 15, 58, 121
spreadsheet, 55, 64, 65, 67, 68, 99
sprint, 32, 33, 55, 65, 86, 90, 100, 104, 105, 106, 107, 109, 112, 113, 114
stakeholder, 7, 33, 43, 52, 56, 66, 67, 84, 92, 98
standards, 34, 81

state, 29, 58, 59, 70, 90, 124, 131, 133
statement of work, 56, 101
step, 2, 11, 18, 22, 23, 27, 28, 29, 31, 33, 34, 48, 52, 55, 56, 66, 71, 72, 73, 74, 78, 84, 89, 106, 124, 125
step-by-step conversation, 22
storyboard, 25, 76, 77
structured, 19, 22, 33, 68, 128
structured text, 19, 22
subject matter experts (SMEs), 39
SUD, 11, 133
sunny day scenario, 11, 22, 133
support, 16, 114, 118
support analyst, 14
swim lane, 26, 27
system behavior, 13, 19, 26
system boundary, 21, 22
system integration, 37, 38, 43, 114
System Under Discussion, 11, 133
system-oriented, 51
systems analyst, 14
task flow diagram, 133
technical specifications, 112
technical story, 105, 106, 112
technical writer, 16, 43, 45
template, 55, 68
test, 30, 33, 51, 102, 114, 132, 134
test case, 2, 3, 16, 31, 64, 75, 133
test environment, 106

Index

test script, 45, 53, 129
test strategy, 45
test-driven development, 3
testers, 2
text, viii, 19, 22, 24, 61, 68, 123, 124, 125, 128, 135
tool, 36, 47, 55, 56, 57, 59, 64, 65, 67, 68, 72, 73, 74, 78, 86, 89, 119
trade-off, 15, 44, 115
training materials, 2, 45
trigger, 70, 124, 134
TW, 45
UAT, 114, 128, 134
UI, 45, 71, 76, 77, 118, 126, 134
UML Activity Diagram, 26
UML Sequence Diagrams, 25
uncertainty, 51, 53, 98
Unified Modeling Language, 21
unique identifier, 63
unit test, 134
usability, 15, 45, 80
usability analyst, 7, 75
use case, 1, iii, iv, vii, viii, 1, 2, 3, 4, 5, 6, 7, 9, 10, 11, 12, 13, 14, 15, 16, 18, 19, 20, 21, 22, 23, 24, 25, 26, 28, 29, 30, 31, 32, 33, 34, 35, 36, 37, 38, 39, 41, 42, 43, 44, 45, 48, 49, 51, 52, 53, 54, 55, 56, 57, 58, 60, 61, 62, 63, 64, 65, 66, 67, 68, 69, 70, 71, 72, 73, 74, 75, 76, 77, 78, 79, 80, 82, 83, 84, 85, 86, 88, 89, 90, 91, 92, 93, 94, 95, 97, 98, 99, 100, 101, 102, 104, 105, 106, 107, 109, 110, 111, 112, 113, 116, 117, 118, 119, 121, 122, 123, 124, 125, 127, 128, 129, 130, 131, 132, 133, 134
use case analysis, viii, 2, 3, 5, 6, 7, 13, 14, 16, 19, 20, 28, 29, 33, 34, 35, 37, 38, 39, 44, 45, 48, 49, 79, 119, 128
use case register, 55, 62, 63, 67, 68, 71
Use Cases
   benefits, 3
   diagram, 21
   discovery, 1, 55
   extend, 22
   fully dressed, 3, 22
   include, 14, 18, 22, 25, 34, 36, 56, 59, 61, 64, 67, 73, 81, 85, 101, 105, 107, 109, 111, 113, 114, 117
   initial, 55, 65, 66, 85, 89
   library, 10, 32, 104
   model, 11, 21
   narrative, 20
   relationships among, 22
   Rosetta Stone, iii, iv, vii, 3, 14, 39
user acceptance test, 114
user analysis, 5, 56, 57, 60, 64, 111
user community, 10, 14, 56, 61, 134
user experience, 45
user interface, 2, 7, 14, 15, 24, 28, 45, 57, 74, 75, 76, 102, 111, 135

user stories, 4, 29, 30, 31, 32, 33, 34, 89, 90, 105, 106, 109, 112, 114
user story, 30, 89, 93, 100, 106, 109, 113, 119
user story mapping, 33
user-centric, 12, 28
user-oriented, 51, 80
UX, 45, 75
validate, 67
validation, 53
value, vii, 7, 8, 22, 29, 34, 36, 39, 114, 115, 119, 132
velocity, 113
vendor software, 37, 38
vertical slice, 30
vision, 49, 50, 57

voice, 12, 23, 28, 38, 43, 59, 74, 92, 129
Voice of the User, 12
waterfall, 3, 5, 85, 97, 101, 102, 103, 104, 135
WBS, 48, 110, 135
web application, 9, 37
website, 36, 37
Widrig, Don, 13, 21, 122
wiki, iv, 64, 65, 66, 68, 135
wireframe, 2, 28, 45, 76, 107, 109, 110, 111, 113, 126, 135
work breakdown structure, 48, 110, 136
workbook, 64, 67
workflow, 25, 27, 28, 109
workshop, 1, 2, 16, 39, 43, 66, 71, 75, 104, 109, 111, 112

About the Author

David A. Bly is widely respected for his abilities in Software Project Management. With 30 years of experience, he has successfully managed software projects for many different companies across several industries in the United States, the United Kingdom, and in Europe. He has also held senior executive positions in the field of project management such as Director of Projects and Chief Project Officer. Dave currently resides in Issaquah, Washington and works as an independent consultant in project management and software engineering practices.

www.ingramcontent.com/pod-product-compliance
Lightning Source LLC
Chambersburg PA
CBHW051807170526
45167CB00005B/1912